Undergraduate Lecture Notes in Physics

Series Editors

Neil Ashby, University of Colorado, Boulder, CO, USA

William Brantley, Department of Physics, Furman University, Greenville, SC, USA

Matthew Deady, Physics Program, Bard College, Annandale-on-Hudson, NY, USA

Michael Fowler, Department of Physics, University of Virginia, Charlottesville, VA, USA

Michael Inglis, Department of Physical Sciences, SUNY Suffolk County Community College, Selden, NY, USA

T0172096

Undergraduate Lecture Notes in Physics (ULNP) publishes authoritative texts covering topics throughout pure and applied physics. Each title in the series is suitable as a basis for undergraduate instruction, typically containing practice problems, worked examples, chapter summaries, and suggestions for further reading.

ULNP titles must provide at least one of the following:

- An exceptionally clear and concise treatment of a standard undergraduate subject.
- A solid undergraduate-level introduction to a graduate, advanced, or non-standard subject.
- A novel perspective or an unusual approach to teaching a subject.

ULNP especially encourages new, original, and idiosyncratic approaches to physics teaching at the undergraduate level.

The purpose of ULNP is to provide intriguing, absorbing books that will continue to be the reader's preferred reference throughout their academic career.

More information about this series at http://www.springer.com/series/8917

Mark Gallaway

An Introduction to Observational Astrophysics

Second Edition

 Springer

Mark Gallaway
Hockley, Essex, UK

ISSN 2192-4791   ISSN 2192-4805   (electronic)
Undergraduate Lecture Notes in Physics
ISBN 978-3-030-43550-9   ISBN 978-3-030-43551-6   (eBook)
https://doi.org/10.1007/978-3-030-43551-6

This Springer imprint is published by the registered company Springer Nature Switzerland AG
The registered company address is: Gewerbestrasse 11, 6330 Cham, Switzerland

*This book is dedicated to Cordi Scott,
Dr. Heather Campbell,
and Anna Christodoulou,
whose dedication to the promotion
of science to the public is inspirational.*

*It is also dedicated to Rachel Tungate,
whose work with Young Carers gives support
and hope to children who deserve a better life
than the one given.*

*Thank you to all of you for making the world
a better place.*

Acknowledgements

Many thanks to Edward Luff, whose help was essential in producing this edition, and to my wife, Sarah, for understanding and tea.

I also wish to thank my former MSc student Peter Beck. Peter continued to do research on ultraprecision photometry despite the fact that he was terminally ill. I always enjoyed his company and our discussions on all aspects of science. This edition would not have been possible without the work he undertook.

Contents

Acronyms

2MASS	Two Micron All Sky Survey: An infrared astronomical survey
AAVSO	American Association of Variable Star Observers: A group of variable star observers and also the catalogue they have created
ADC	Analogue-to-digital converter: A device to convert an analogue signal such as voltage to a digital one such as count
ADS	Aitken double star catalogue
Alt	Altitude: the vertical position of a telescope
AO	Adaptive Optics: A process using movable optics to improve optical resolution
APT	Aperture photometry tool: A freeware multiplatform photometry tool
AU	Astronomical unit: The mean distance between the Earth and the Sun, most often used in solar system physics
Az	Azimuth: The horizontal position of a telescope
CCD	Charge-coupled device: The detector in a modern astronomical digital camera
CDS	Centre de données astronomiques de Strasbourg: The French body that hosts SIMBAD
CMB	Cosmic microwave background: The afterglow, at microwave wavelengths, of the light that escapes during the era of recombination
CMD	Colour-magnitude diagram: A plot of absolute magnitude against colour. The Hertzsprung–Russell diagram is an example of a colour–magnitude diagram
CMOS	Complementary metal oxide semiconductor: A photon detector similar to the CCD
CTE	Charge transfer efficiency: The rate at which electrons are lost during pixel-to-pixel transfer in a CCD
CTIO	Cerro Tololo Inter-American Observatory: An international observatory based in Chile

CYMK	Cyan, yellow, magenta, black: A colour scheme for colour printing
Dec	Declination: The vertical location of an object on the celestial sphere
DSS	Deep Sky Stacker: An application for stacking frames
DSS	Digital Sky Survey: A catalogue made from digitised optical Schmidt plates
E-ELT	European Extremely Large Telescope: A 39 m diameter telescope currently under construction in Chile
ESA	European Space Agency: The European Space body
FFT	Fast Fourier transform: A mathematical process used to apply a Fourier transform
FITS	Flexible Image Transport System: The standard format for astronomical images
FOV	Field of view: The angular diameter of the area of the sky that can be seen by a telescope
FWHM	Full width half maximum. The width of a curve measured form half its maximum value. In photometry, this is related to the seeing
GMT	Greenwich Mean Time: The current time for an observer on the Greenwich meridian line, without adjustments for daylight saving
GPS	Global Positioning System: A satellite positioning system operated by the US government
GSC	Hubble Guide Star Catalogue. A large catalogue of bright stars used for the point of the Hubble Space Telescope
GVCS	General Catalogue of Variable Stars: A catalogue of variable stars
HD	Henry Draper Catalogue: A stellar catalogue
HDC	Henry Draper Catalogue: An HD supplement
HDEC	Henry Draper Catalogue: An HD supplement
HI	Hydrogen 1: Monatomic hydrogen
HII	Hydrogen 2: Ionised hydrogen
HR	Hertzsprung–Russell: As in the Hertzsprung–Russell diagram
HST	Hubble Space Telescope: The first large space-borne optical telescope
IAU	International Astronomical Union: The international governing body for astronomy
IC	Index Catalogues: A supplement to the NGC
IFU	Integrated field unit: A device that allows whole-field spectrography
IRPC	Infrared Processing Centre: NASA science center tasked for data processing, analysis, and archiving of NASA's infrared data
IRSA	Infrared Science Archive: An online database integrating various infrared astronomical catalogues
ISO	Infrared Space Observatory: A space-borne infrared telescope
Kpc	Kiloparsec: 1000 parsecs

LIDAR   Light detection and ranging

LRGB   Luminance, red, green, blue: A colour image with a luminance frame

LST   Local Sidereal Time: The current time at an observer's location based
            on the transit of astronomical objects across the local meridian

MOS   Multi-object spectrography. A device for taking many spectra at once

Mpc   Megaparsec: 1 million parsecs

MPC   Minor Planet Center: The IAU body responsible for solar system
            objects. The MPC holds an extensive database of comets and asteroids

NDF   Neutral density filter: A filter that reduces brightness without affecting
            the spectral ratio of the source

NGC   New General Catalogue of Nebulae and Clusters of Stars: A large
            astronomical catalogue of nonstellar sources

NOFS   Naval Observatory Flagstaff Station: A large observatory located in
            Arizona

NOMAD   Naval Observatory Merged Astrometric Dataset: A large astronomical
            catalogue produced by, amongst others, the NOFS

pc   Parsec: A unit of astronomical distance equating to approximately
            3×10^{16} m. The parsec is the distance of an object that would have a 1
            arc second parallax

PNe   Planetary nebula: A type of nebula-related formation of a white dwarf

PSF   Point spread function: In photometry, the function that describes how
            the light from an object spreads over the frame. It is most often a
            Gaussian

QE   Quantum efficiency: The efficiency of a quantum device, for example,
            a CCD

RA   Right ascension: The horizontal location of an object on the celestial
            sphere

RBI   Residual bulk image: The appearance of a false star in an image caused
            by the saturation of a CCD

RGB   Red, green, blue: A colour image

RMS   Root mean square: The square root of the mean of the squares of the
            sample

SIMBAD   Set of Identifications, Measurements, and Bibliography for
            Astronomical Data: an online searchable database of astronomical
            objects

SNR   Signal-to-noise ratio: The ratio of the signal to the noise of the system.
            A general indication of uncertainty

STScI   Space Telescope Science Institute: The organisation that manages the
            day-to-day operation of the Hubble Space Telescope

UGC   Uppsala General Catalogue: A northern hemisphere catalogue of
            galaxies

UT1	Coordinated Universal Time One: A more modern version of GMT. Effectively, there is currently little difference between UT1 and GMT
WCS	World Coordinate System: The standard translation format for plate location to sky location translation
WDS	Washington Double Star Catalogue
WISE	Wide-field Infrared Survey Explorer: An infrared space-borne telescope
Z	Cosmological redshift: The ratio between the wavelengths of a spectral line

Chapter 1
Introduction

Abstract During the course of your astrophysics degree, you will be undertaking observations of the skies. This book is designed to guide you through observational astrophysics through a series of practicals. You will learn important techniques in photometry, spectrography, astrometry, and general imaging as well as being introduced to some of astronomy's more arcane features.

1.1 The Reader

Most undergraduate astrophysics degrees have a practical component. This practical component is usually undertaken using a small telescope, on campus if you are lucky, or otherwise at a large off-site facility. Remote observations are also sometimes used, carried out via an internet link between a local computer and a robotic telescope. The amount of time you might spend at an instrument can be surprisingly short. Clearly, this is not an ideal situation, as observational astronomy is a complicated subject and one that has developed over many centuries.

If you do not feel ready for the task that is in front of you, you are not alone. Scientific observation is a complex and somewhat daunting task, particularly when you find yourself alone in the dark late at night. This book was written to help new undergraduates studying astrophysics. It is intended to serve as a primer to be consulted prior to observing, and also as a handy reference during the observing process itself. The book can also be referred to later when you are busy processing the images you took. However, a word of caution—this book will not act as a substitute for thorough note-taking. It is always a good idea to document what you are doing while you do it and while it is fresh in your mind. Also, with regard to the sections on data reduction and image analysis, the volume you are holding is not specific to any single observatory. Rather, it is intended as a general reference. Nor is it intended to be a complete reference for any of the areas of astronomy covered in the following chapters, each of which could, in itself, be a topic for a complete book.

While this book is aimed at undergraduates, I hope that it will also be useful for postgraduates, particularly for those who do not have much direct experience

© Springer Nature Switzerland AG 2020
M. Gallaway, *An Introduction to Observational Astrophysics*,
Undergraduate Lecture Notes in Physics,
https://doi.org/10.1007/978-3-030-43551-6_1

of observational astronomy or are out of practice. It should also be useful for the amateur astronomer who wishes to do some scientific work with a home instrument. One exciting aspect of astronomy is that it is a modern science in which even an amateur can do highly useful research.

1.2 Required Equipment and Software

For the purposes of this book, it's assumed that you have access to a permanently mounted and fixed telescope. We further assume that it is a Schmidt–Cassegrain telescope (we'll discuss specifically what this is later on). The most widely known manufacturers of small telescopes are Meade and Celestron, whose telescopes are the most commonly used by universities. (The make of the telescope affects only a small portion of this book, so don't worry if yours is different!)

We're also going to assume that the telescope has a mount that is capable of sidereal tracking, and also that it has an astronomical digital camera. The camera should be able to provide guidance information to the mount. Also, you will need scientific-grade astronomical filters (ideally, these will be contained in the filter wheel that is part of the camera).

Your telescope may be controlled by either a computer or a hand controller. The camera will always be computer-controlled. There is a range of different software applications for controlling camera and telescope; the most popular are ccdSoft and MaxImDL. These products are broadly similar, although for the purposes of this book, we will be using MaxImDL. In addition, you may also find that your telescope has a powered focuser. This tool isn't essential, but it is a useful add-on.

Throughout this book, you will be asked to manipulate the images you take with your camera/telescope combination. Both ccdSoft and MaxImDL are capable packages, but both have associated licensing fees. There are also freeware image manipulation applications available for Windows, OS X, and Linux. In this book, we will be using SAO DS9, Aperture Photometer Tool (APT), AstroImageJ, and SALSAJ. Some universities may require you to use certain professional-grade applications. A common example of such a tool, IRAF, is versatile and fast, but is designed only for Linux (which also extends to OS X in practice). IRAF requires the use of a command-line interface and is notorious for its steep learning curve, so you may find that to be something of a speed bump when you start out. However, one of IRAF's merits is that it can be scripted to run a series of linked commands. This can greatly speed up many image analysis processes. Most of the image manipulation that this book will cover can be done in IRAF if you wish.

This book also covers tasks that can be carried out using commands in the Python programming language. Python is becoming increasingly popular in professional astrophysics, and I can't recommend it highly enough as a first language. I also highly recommend learning a programming language, as it will help you in your course and also for getting a job post-degree. Python is open source (i.e., it's free), multi-platform, well supported, and easy to check for errors. It has many powerful add-on packages called modules, which include those for performing astronomical calculations, astronomical image manipulation, statistics, and even a package to control IRAF via python (PYRAF). A number of Python distributions and versions are currently in use. The author's preference is the Anaconda Python 2.7.6 distribution, as it is easy to install and comes with most of the tools you need. However, others are available and might already be installed on your computer.

This book also covers spectrographs and radio observation. Most universities running astrophysics programs have telescopes with cameras, but having a mounted spectrograph is rarer, and access to a radio telescope even more so. However, please feel free to read these chapters even if you don't currently have access to such instruments. These are, after all, important tools in professional astronomy, and you never know when this knowledge might prove useful.

1.3 How to Use This Book

This book doesn't have questions for the reader to attempt, although it does provide worked examples for the trickier mathematics. Instead, there are practical activities for you to try. The idea is that you will gain the skills and the confidence you need by actually having a go at these things and seeing how they work in practice.

The initial chapters cover the underpinnings for the practical sessions introduced later. In most cases, it's possible to do the exercises from previously acquired data. (This is sometimes referred to as "canned data.") This does, however, provide a different experience from actually going out and doing the observations. (And don't underestimate how different sitting indoors in front of a computer in the mid-afternoon feels from actually being out there on a cold winter night.)

Unlike many books that you may encounter during your studies, this one is not limited to a single course or module. Rather, it's intended to be a companion that can guide you through all the stages of your course. Topics we will cover include general observing, solar system astronomy, solar physics, planetary science, stellar astrophysics, extragalactic astrophysics, and cosmology. We will also touch on broader physical topics such as quantum mechanics (where relevant to the matter at hand).

The practical exercises are laid out in the order that I teach them in my own classes. Your course may have a different structure; if so, please feel free to skip backward and forward as you need to. However, keep in mind that the initial chapters and practicals are vital, as they cover essential observational skills. Even if you are not taught them, you should make yourself familiar with the materials and techniques therein. Whilst reading this book, you will likely also notice that some ideas are repeated between chapters. This is both to reinforce key concepts and also to allow the reader to skip forward to a desired section without the need to read every single item in between.

Your course likely requires a final-year project. By the end of this book, you should have the needed skills and confidence to take this on and succeed. This book will help you in deciding on a project and its structure.

Astronomy is arguably the oldest of the sciences—the positions of the stars, the Moon, and planets have been used for thousands of years to determine times for the sowing and harvesting of crops and for navigating the seas. The temporal power of the Mayan priesthood rested partly on their ability to reliably predict eclipses, and many ancient monuments across the world feature astronomical alignments as part of their design. Astronomy's prehistory segues into what we would now consider astrology, just as chemistry has its origins in alchemy. Astronomy was clearly established as an actual, recognisable science only as recently as the seventeenth century. Sir Isaac Newton's formulation of the theory of gravity was an essential achievement, as it provided a systematic theoretical foundation to what had previously been something of a disorganised grab-bag of facts, half-truths, and conjecture. Because of this extensive pre-history, modern astronomy contains many odd and quirky traditions (as an example, the sexagesimal coordinate system originated with the Babylonians). You may well find some of these oddities rather disorienting at first. However, you will get used to such quirks in time.

Throughout this book, we have preferentially used the SI unit system. There are some deviations from this rule, as astronomy has plenty of non-SI measures, such as, for instance, the parsec, the jansky, and the astronomical unit. Where there is a suitable SI unit, though, we will be sticking to that, so you won't be bedevilled by angstroms or ergs. You will find Joules and nanometres, though. If you need to use a non-SI unit, please remember to make the appropriate conversion.

Before we move on, let's briefly define a couple of these unavoidable non-SI units. The astronomical unit (AU) is the median separation between the Earth and the Sun. It's equal to 149,597,870,700 m. The AU is most often used in dealing with distances inside the solar system. The Parsec (pc) is the distance at which an object would show a parallax displacement of one second of arc. It equates to $3.08567758 \times 10^{16}$ m. Kiloparsecs (Kpc; 1 Kpc = 1,000 pc) and megaparsecs (Mpc; 1 Mpc = 1,000,000 pc) are also often used in astronomy. For the most distant observable objects, the distance is expressed in terms of the degree of observable cosmological redshift (z). Higher-z objects are further away, but the scale isn't linear, and moreover, the actual distance will depend on the cosmological model being used. As a point of reference, objects at $z = 1$ are roughly halfway to the edge of the observable universe, whereas the cosmic microwave background is at $z = 2,000$. (The CMB effectively constitutes the "edge" of the observable universe for all practical purposes.)

1.4   Introduction: Best Practice

The observatory that you will be using is a laboratory, just like any other that you may have worked in. As such, you should treat it as a lab—read and understand the safety guidelines and the operating instructions before you begin observing. It is very likely that there will be specific documentation relating to the use of the observatory. You need to make yourself aware of what is in those documents, and if you have any questions, it's a good idea to ask a member of the observatory staff.

Prepare properly for a night's observing. Before you set off, you need to think about what you're planning to do and how you will go about achieving it. First of all, if your observation is part of a course, then make sure that you've read the practical notes and that you understand them. You should have a list of the observational targets that you want to work on and their coordinates in the sky. You should also know what exposure times and filters you need. You should also note the position of the Moon in the sky. This can become important, especially if it's at or near full phase.

If it's possible to find out in advance, it's a good idea to know what instrument and camera you will be working with. Not all cameras and instruments are equal, and some specific setups may have implications for your work. It's also a good idea to have a list of backup objects in case your primary targets are obscured. Never assume that there won't be a single solitary cloud parked in the worst possible bit of the sky. Telescope time is expensive, and the weather is not dependable.

It's also important to have some means to get your images off the camera control computer, such as memory sticks and cloud services. Different observatories have different standard methods, so check with the staff.

If you do as much preparation as you can in advance, you will have a rewarding and reasonably stress-free observing session.

The most important tool you will have with you whilst observing is your lab book. You need to record everything that happens during the course of your observations. Include the telescope name and its optical characteristics (we'll return to these details later on). Here is a list of the things you need to record:

- the time and date of each observation;
- the instrument you're using on the telescope;
- the exposure time for each observation;
- the filter used for each observation;
- the ID of the target that's being observed;
- its location on the sky.

In addition, you should record the weather conditions, the sky brightness (if known), and the seeing (i.e., the quality of telescopic observation). We'll show you how to work this out later on. These details are essential if you later wish to make a quantitative assessment of your data quality. If you are observing as part of a group, don't assume that someone else is taking all the notes. They probably are not. Always make your own notes and make them as thorough and detailed as possible. Keep in mind that precision is a key aspect of experimental science. If at all possible, always record the uncertainties on the data that you're taking. You should ask yourself the question, "Could another scientist reproduce my methods based on what I have written down?" The answer should always be yes.

Given the nocturnal nature of astronomy, it can be extremely cold when you are observing, so ensure that you dress appropriately for the weather, including footwear. It is easy to remove a jacket if you are too warm, but you cannot put one on that you have not brought. Telescope domes are unheated, so unless you are lucky enough to have a control room, you are as likely as not going to be exposed to the elements.

You are also going to be stationary for considerable amounts of time. Consequently, hypothermia can become a problem. If you feel too cold, go somewhere warm for a while.

You will be observing in a dark environment for all but solar and radio observations. Initially, it may appear extremely dark, however after a while, you will become dark adapted, and it will become easier to see. It takes about 20 min to become fully dark-adapted, but only a few seconds of exposure to bright light to cause you to lose your adaptation. Remember this when entering lit buildings and when using a torch or flashlight. If you flash it in another observer's eyes, they will lose their dark adaption, or stray light may fall into the instrument, ruining the observation. Exposure only to red light reduces the loss of dark adaption, although it still has an impact. If your observatory has red lighting, try to use it. Likewise, if you have access to red torches, please use those.

1.5 Robotic and Remote Observing

In recent years, both robotic and remote observing have become increasingly popular, as they offer the chance to use a telescope, often in remote locations, without the inconvenience and cost of extensive travel. It also reduces the cost of having somebody look after the telescope during the night and the day. It may come as a surprise, but many of the servicing tasks for telescopes have to be done in daylight. This means that astronomers now have access to dozens if not hundreds of small telescopes that were previously not worth using due to the cost involved. This comes at a time when long photometric observations have become increasingly important in the search for exoplanets and supernovae, as well as for potentially Earth-crossing asteroids, tasks for which small robotic telescopes are ideal. Examples of such robotic telescopes are the two-meter Liverpool Telescope in the Canaries, the Open University's PIRATE telescope, and TRAPPIST, at the La Silla Observatory, in Chile.

Robotic telescopes are unmanned instruments that use hardware automation and sophisticated software to undertake astronomical observations. They require some kind of weather station to determine whether the weather conditions meet the criteria for the observation. Typically, this will be accomplished with an IR sky sensor that measures the temperature of the sky. A warm sky is cloudy, while a cold sky is clear. A robotic sight might also use an all-sky camera attached to image-processing software that can determine cloud coverage and location. In both of these cases, it is almost impossible to detect high clouds, which can badly effect high-precision photometry observations. Observatories can overcome this problem using LIDAR, but that is rare. With robotic observing, the user submits a job to the queue. This is normally done via some kind of user interface that will filter out some of the dumber mistakes people make. Most often than not, this job is then reviewed by somebody on the observatory's time-allocation team, and if accepted, the job is given a priority based on a number of factors and then passed to the telescope. Once in the telescope queue, it will become available to run. Most robotic instruments use an algorithm to determine which observation to run next. Obviously, it will filter out any jobs whose target is not visible for some or all of the night, it will then look at the sky conditions and determine which observations have condition limits that are not met by the current conditions, and filter those out. It will then look at the priority of each unfiltered job and determine which one to run.

Remote telescope operation is slightly different from robotic telescope operation. Remote telescopes are telescopes whose control computer can be controlled remotely via, for example, VNC or Windows Remote Desktop. With remote telescopes, you are effectively running the telescope; you are just not in the dome. All robotic telescopes are capable of remote operation, but not all remote telescopes are capable of independent operation; they may lack the necessary software (and some hardware). There is no queue, just you, a computer and a telescope, possibly thousands of kilometres away. If you have access to a robotic telescope and have a time-critical observation, then it might be worth asking whether you can use the telescope in remote mode rather than the robotic mode. Then you can be sure, barring bad weather or equipment failure, that your observations will be made. Of course, the downside of this is the need to be at the computer in the middle of the night, whereas if you were in the robotic mode, you could be home in a nice warm bed.

One last operational mode we should discuss is service mode. You are unlikely to encounter this as an undergraduate, but you will encounter it quite a bit if you go on to a career in observational astrophysics. In service mode, you write a telescope plan, and a person in the observatory takes the observation for you without you having to be there. Many very large telescopes are heavily automated, but the operational cost of a system failure is too high to run in robotic mode. Telescopes like the Very Large Telescope in Chile cost in excess of €10,000 per night to run and are heavily oversubscribed. Robotic telescopes have a failure rate that is currently unacceptable to these big observatories, and service mode is a halfway house between robotic observations and actually flying out and doing the observation yourself (which despite being really exciting the first few times is costly and not good for the environment).

Chapter 2
The Nature of Light

Abstract Our understanding of the universe at large comes almost exclusively from our understanding of light and how quantum mechanics explains the physical features we see in the spectra of astronomical objects. Astronomers measure the brightness of astronomical objects in terms of flux and magnitude. The use of coloured filters enables us to examine specific characteristics of the object being observed.

2.1 Introduction

The physics of light is central to our understanding of astrophysics. For the vast bulk of history, astronomy was confined to the so-called visible spectrum, that wavelength region that can be seen by the human eye. In the year 1800, William Herschel—famous for the discovery of Uranus, the first new planetary discovery since prehistory—attempted to measure the temperatures of individual colours in the rainbow. Those efforts led to an inadvertent discovery. Herschel's thermometer continued to detect warmth even after it was moved outside the visible spectrum. Herschel had accidentally discovered the existence of infrared light.

Infrared (IR) astronomy began to develop as a quantitative science in the mid-1800s (the first actual object detected in the IR was a cow, as it happens).

Radio astronomy began as a discipline in the 1940s, having been made possible by advances in electronics and electrical technology. Subsequent to this, ultraviolet (UV), X-ray, millimetre and submillimetre observations also became possible. Many of these depend on space-based or balloon-borne instruments, due to the opacity of Earth's atmosphere at many wavelengths. (This is also why observatories tend to be built at high altitude—the more of the air you're above, the better. Water vapour in particular can bedevil infrared astronomy, due to its deep absorption features, and mid-infrared astronomy is effectively impossible at sea level.)

Classical physics presents light as a travelling wave consisting of an oscillating electric field with a perpendicularly oscillating magnetic field. Both the electric and

© Springer Nature Switzerland AG 2020

M. Gallaway, *An Introduction to Observational Astrophysics*,
Undergraduate Lecture Notes in Physics,
https://doi.org/10.1007/978-3-030-43551-6_2

magnetic fields are at right angles to the direction of travel. The energy of the wave is inversely proportional to its wavelength, with longer waves having lower energies and shorter waves having higher energies. (In terms of visible light, red light has a longer wavelength than blue light, and so carries less energy.)

The relationship between frequency and wavelength is given by (2.1), where c is the speed of light ν is frequency, and λ is the wavelength:

$$c = \nu\lambda \tag{2.1}$$

Hence as wavelength increases, frequency decreases.

Conversely, in quantum mechanics, light can be considered to be a particle, the photon, with energy given by (2.2), where the energy is E, h is Planck's constant, frequency is ν, and wavelength is λ:

$$E = h\nu = \frac{h\lambda}{c} \tag{2.2}$$

It is important to realise that light is neither a wave nor a particle; rather it is both simultaneously, and it is only the method of observation that determines whether it has the features of a wave or a particle. Hence when, for example, when you are looking at light's interaction with optical instruments, its behaviour is wavelike, but within our cameras, it behaves like a particle.

Within the research community, there is a tendency for different groups to have each its own idiosyncratic means to describe the light they work with. Those who work at short wavelengths (such as X-ray astronomers) tend to characterise their photons by their energies. (This approach is also used with cosmic rays.) Observers in the UV to IR range tend to refer to their light by wavelength. Lastly, those in the millimetre and radio regimes tend to describe things in terms of frequency.

The electromagnetic spectrum itself is continuous. For human convenience, we divide it up into a number of bands, although the exact cutoff locations for the bands are disputed. Table 2.1 shows these various bands with their typical energies, wavelengths, and corresponding frequencies.

Table 2.1 Table showing the various bands of the electromagnetic spectrum and their approximate wavelength, frequency, and energy

Name	λ nm	ν Hz	E eV
Radio	$>10^8$	3×10^8	10^{-8}
Microwave	$10^8–10^5$	$3 \times 10^9–10^{12}$	$10^{-5}–0.01$
IR	$10^5–700$	$3 \times 10^{12}–4.3 \times 10^{14}$	$0.01–2$
Visible	$7{,}000–400$	$4 \times 10^{14}–7.5 \times 10^{15}$	$2–3$
UV	$400–1$	$7.5 \times 10^{14}–3 \times 10^{17}$	$3 – 10^3$
X-ray	$1–0.01$	$3 \times 10^{17}–3 \times 10^{19}$	$10^3–10^5$
Gamma ray	<0.01	$>3 \times 10^{17}$	$>10^5$

2.2 The Quantum Nature of Light

Photons are emitted when there is a change in the energy state of a system. For example, the model often used when the atom is introduced to students is the idea of the electron orbiting the nucleus. Although this is strictly not true in all situations, it is a good enough model for most purposes. Electrons that are orbiting close to the nucleus have less energy than those orbiting farther away. Hence when an electron gains sufficient energy, it moves to a higher orbit, and when an electron emits a photon, losing energy, it moves to a lower one.

In the late 1800s, Heinrich Hertz studied the photoelectric effect. He noticed that certain materials would emit electrons when exposed to light—but the release occurred only when they were exposed to certain wavelengths. If the wavelength was wrong, it didn't matter how intense the light was. No current would be observed. This was something of a puzzle—how did the materials "know" that they were being exposed to the "wrong" sort of light? Why would they seem to care about anything except luminous intensity?

It was Albert Einstein who suggested a solution (and it was actually this solution that earned him his Nobel Prize). He suggested that light is quantised, that is, that its energy is delivered in discrete packets. These packets are, of course, the particles that we call photons. Also, Einstein theorised that instead of there being a uniform continuum of energy levels in which electrons can exist, there are actually just a few discrete ones—that is, electron energy levels are quantised too. For an electron to move between these fixed levels, an incoming or outgoing photon has to have just the right energy. If it is wrong, then the photon cannot absorb it and thus will not respond. This is why the materials in Hertz's experiments reacted only to specific wavelengths of light.

As to why electrons display this behaviour, imagine them for a moment not as particles, but rather as waves that are wrapped around an atomic nucleus. For a wave to wrap neatly, its orbital circumference has to correspond to a multiple of its wavelength. If the circumference doesn't follow this pattern, then some of the wave's peaks and troughs will overlap each other, and the wave will interfere with itself. Applying this to an electron, we can see that a situation would arise in which an electron could interfere itself out of existence, thus strongly violating the principle of conservation of energy. This has the effect of forbidding electrons from occupying such orbits.

These insights, along with others, led to the emergence of quantum mechanics. It is worth noting that although quantum mechanics has many weird aspects, it is nonetheless one of the most successful and important scientific theories ever developed. It underlies not just the multibillion-dollar fields of electronics and computing but has practical applications throughout the whole of industry and commerce. It even has implications for seemingly unrelated fields such as biology and medicine, through applications such as imaging machines and analyses of protein folding. Whilst quantum theory is certainly weird, it is a necessary weirdness.

The key insight of quantum theory is that the quantisation of energy in a bound system applies to every change between two states. Hence not only is the energy of electrons quantised, so are the spins, the momenta, and also molecular vibrations and rotations. There are some conditions under which this rule breaks down, of course. The emission caused by the deceleration of electrons in a magnetic field is not quantised (note that we are considering free electrons here, moving outside an atomic shell), and the emission caused by the capture of electrons by an ion is also nonquantised. (Note again that the electron starts off free rather than bound—this is what makes all the difference.) The emission of photons due to the capture of electrons by atoms is called free-bound emission.

When the idea is applied to processes within our Sun, quantisation leads to an important realisation. The Sun's core temperature is extremely high, around 15 million K. As a result, the material inside it is in a highly ionised state—the Sun is composed of plasma. This means that most emission is of the free-bound variety, so the Sun should exhibit a continuous emission spectrum.

But it doesn't. The Sun actually has some bright emission lines such as the Balmer and Lyman series.

What is going on? The answer lies in the fact that whilst the emission of photons has no dependency on direction, the absorption of photons does have a directional bias. As photons near the star's surface, there is more star below them than above, so by this point, when a photon is absorbed, that photon is most likely from somewhere deeper inside the star. This causes dark features known as absorption lines to appear in the star's spectrum.

But also, given that a star's surface is much cooler and less dense than its core, it is possible for some bound absorption to occur at or near the surface. This will show up as bright emission lines, such as the aforementioned Balmer and Lyman features. The Balmer lines are caused by the transition of a hydrogen electron down to the level-2 energy state. The first Balmer line, called the H-α line, is seen at $\lambda = 656.3$ nm (in the red area of the visual spectrum) and is caused by a 3-to-2 transition. The second Balmer line, H-β represents the 4-to-2 transition, and so on through the other transitions.

The Lyman series covers the transitions down to the $n = 1$ energy level. The first, the 2-to-1 transition, is called the Lyman α line.

When seen under laboratory conditions, an emission line will be narrow. What broadening there is will be due partially to the optics used to observe it, and partially to the uncertainty principle. However, actual stars are less tidy than well-organised laboratories. Consequently, the thermal motion of the plasma within the star will spread the lines out due to the Doppler effect. In addition, the bulk motion of the plasma and also the motion of the target with respect to the observer (rotation, radial velocity, etc.) will shift the line slightly toward either the blue or red end of the spectrum. However, these effects don't broaden the lines.

The quantisation of light is the underlying physics for the science of spectrography. In Chaps. 14 and 15 we will go into this subject in much greater depth.

2.3 Measuring Light

You will often come across the terms **luminosity** and **Brightness**. The luminosity of an object is the total amount of energy that is emitted across a specific wavelength range. The brightness is a related but subtly different concept: the brightness of an object is the amount of energy it emits through a specific wavelength range and over a solid angle. Brightness is also dependent on the distance from the emitting object and thus will decline in proportion to the square of said distance.

The **absolute magnitude** of an object is the brightness it would have at a fixed reference distance. The point of this is that it allows a quick and valid comparison with the absolute magnitudes of other objects. For solar system objects, the reference distance is 1 AU. objects outside the solar system, the absolute magnitude is the magnitude that the object would have if it were located at a distance of 10 pc.

In general, you should use the term "apparent brightness" to avoid confusion.

You may also come across the term "bolometric luminosity" with respect to brightness. It refers to luminosity or brightness across the entire spectrum.

Stellar spectra broadly approximate those of **black bodies**. That is, they can be modelled as perfect absorbers and emitters, and they reflect next to no light. (A little-known fact about the Sun is that in terms of its reflectivity, it is actually about as black as coal.) This means that their output follows the pattern given by Planck's law

$$B_\lambda(T) = \frac{2hc^2}{\lambda^5} \frac{1}{e^{\frac{hc}{\lambda kT}} - 1}, \tag{2.3}$$

where $B_\lambda(T)$ is the total radiation at wavelength λ and temperature T, h is Planck's constant, k is Boltzmann's constant, and c, as usual, is the speed of light in vacuum.

In general, astronomical bodies emit their light **isotropically**. This means that they emit equally in all directions. There is a small handful of exceptions, such as pulsars and quasars (pulsars emit their radiation in beams). However, the majority of objects that you encounter will be isotropic radiators. How do we analyse isotropic emitters? We can begin by considering a sphere of radius r that surrounds an isotropic emitter. The amount of energy passing through any given area of the sphere will decline as the sphere gets larger; keep in mind that energy is conserved, so as the sphere grows, the same amount of energy will be spread over a larger and larger area. This means that the flux is directly proportional to the emitter's luminosity and inversely proportional to the square of the distance between emitter and observer. Hence flux can be defined as follows:

$$F = \frac{L}{4\pi r^2}. \tag{2.4}$$

This implies that the total amount of flux that a telescope's aperture can gather is dependent on the diameter of the telescope. A smaller one will gather less, a bigger one will gather more. If it helps, think of a telescope as acting for photons as a bucket does for rainfall.

The **instrumental flux** is specific to a given instrument and its setup, since it depends on exposure times, aperture sizes, and the optics being used. In order to turn an instrumental flux back into a physical one, corrections have to be made for all of these factors and also for the relevant atmospheric parameters under which the observation occurred. We will return to this in the chapter on photometry.

2.4 The Magnitude Scale

Instruments typically return their raw observational results in the form of counts per second—that is, the total number of photon strikes on the receiving area divided by the time taken for the observation. (There are some exceptions—radio astronomy uses a non-SI unit called the Jansky, which we will discuss later.) However, in the vast majority of cases, you will be using so-called magnitudes. The magnitude scale is a logarithmic one that has been in use since antiquity.

In the nineteenth century, the magnitude system was quantified and put onto a rigorous, empirical footing. It was discovered that the scale is actually logarithmic, and the scale was calibrated so that a difference of five magnitudes is equivalent to a difference of one hundred in the intensity of flux. Perhaps surprisingly, it turned out that for most stars, Hipparchus's assessments were actually roughly correct.

We now know that Hipparchus's system was actually a bit off. The brightest star in the night sky, Sirius, has a magnitude of -1.46, and 15 other stars are brighter than magnitude 1. Also, the modern system has been extended to include the Sun and the planets. Venus can attain a magnitude of -4.5, the full Moon is around magnitude -13, and the Sun is around -26.7. Given that Hipparchus was working without telescopes or modern equipment, and, in fact, lived before the concept of negative numbers had been developed, we can probably forgive him for missing some details.

As to the logarithmic nature of the system, the most likely explanation for this is that the human eye scales its perceived response to incident light logarithmically rather than linearly. Certainly, the eye can perceive enormously different levels of illumination without returning a sense of difference as great as that which quantitative methods show us are there. However, the whole relationship between physical stimulus and subjective perception remains one of active research in human physiology and psychology.

When the magnitude system was quantified, the British astronomer Norman Pogson introduced the equation that now bears his name. This equation relates the observed magnitudes of two objects to the fluxes they are emitting:

$$m_1 - m_2 = -2.5 \log_{10}(F_1/F_2), \tag{2.5}$$

where m_1 and m_2 are magnitudes of two celestial bodies, and F_1 and F_2 are their corresponding fluxes. Hence, the flux ratio of two stars of magnitude 1 and one of magnitude 6 is $100/1$.

The magnitude of an object is dependent not only on the physical characteristics of that object but also on its distance from the observer, the amount of material between the observer, the detector used, and the transmission characteristics of the instrument, which is often determined by a filter. Filters are discussed at length in Sect. 2.5, but you should be aware that magnitudes should always be quoted as being in a particular filter band. If they are not, then it is assumed that they are unfiltered visual magnitudes. Hence, a star or galaxy may be of magnitude 10 in the R band but of magnitude 12 in the B band. The magnitude of extended and therefore resolved objects is the integrated magnitude, i.e., the combined light over the whole surface of the object, and is known as a **surface brightness**. You should, therefore, allow more exposure time for extended objects than you would for stellar objects of the same listed magnitude.

Most magnitude quotes are **apparent magnitudes**, the brightness as seen from Earth. An important characteristic of a star is its **absolute magnitude**, the brightness of a star at a distance of 10 pc, which hence relates directly to luminosity. Since brightness drops off with the square of the distance, we can modify (2.5) to determine absolute magnitude from apparent magnitude and distance:

$$M = m + 5 - 5\log_{10}(d/pc), \tag{2.6}$$

where M is absolute magnitude, m is apparent magnitude, and d is distance in parsecs.

A major challenge in galactic astronomy is the problem of interstellar reddening. Gas, and in particular the dust between the object being observed and the observer, will preferentially scatter blue light over red. Hence, an object with more scattering appears redder. This would not be much of a problem if the material were distributed uniformly throughout the galaxy (in fact, it would be a major advantage if it were). However, it is not, so consequently, we need to modify (2.7) to adjust for reddening:

$$M = m + 5 - 5\log_{10}(d/pc) - A, \tag{2.7}$$

where A is the reddening in magnitudes.

Two photometric calibration standards are in common use, Vega and AB. The Vega system assumes that the star Vega, α Lyra, is of magnitude zero at all wavebands, which has the advantage that it simplified Pogson's equation to

$$m_2 = -2.5\log_{10}(F_1/F_2), \tag{2.8}$$

so that the magnitude of a star in any filter is just the logarithm of the ratio of its flux over that of Vega (multiplied by -2.5).

The AB system, unlike the Vega system, is based on calibrated flux measurements. For completeness, we give the definition of AB magnitude:

$$m_{AB} = -\frac{5}{8}\log_{10}\left(\frac{f_v}{Jy}\right) + 8.9 \tag{2.9}$$

where f_v is the measured spectral flux density, the rate at which energy transfers through a surface measured per second per unit area. In practice, AB and Vega are almost identical, at least for objects with reasonably flat spectral energy distributions. For the purposes of the practicals in this book, we will always use Vega magnitudes. However, you should always note in your lab book which system you are referring to.

Keep in mind that the flux you measure is the instrumental flux. This means that when you apply Pogson's formula or the various derivatives to that flux, what you will get is an instrumental magnitude. If you wish to determine an actual physical magnitude, then you will need to undertake a much lengthier process, one that accounts for the optics and the specific filter set that you are using.

Another subtlety concerns extended objects such as galaxies and nebulae. The magnitude that is usually quoted for these is the integrated magnitude, which is the total amount of light emitted by the object across the band in question. However, this causes an observational difficulty. If you set your exposure time for such an object based on its integrated magnitude, it will be underexposed, since the light is spread out over more pixels than a star. A workaround for this difficulty is to use the surface brightness, which is the magnitude per square arc second. (Effectively, this is a magnitude density, if that makes sense.) The surface brightness has some issues of its own in that it, too, has inbuilt assumptions—it treats objects as having even brightness, when in fact, most of them will have concentrated brightness in one area and more diffuse brightness elsewhere. It is best to keep in mind what sort of object you are observing and to try to adjust your practice accordingly.

Later in this book we will discuss in depth the concept of **photometry**, the practice of measuring of the amount of light received from an astronomical object over a given time with a given aperture and converting that into magnitudes. Within photometry there is an important concept, that of the **zero point**, which is the magnitude of an object that for the specified instrument setup, will produce one count per second. More typically it is expressed as a flux or count, so that the magnitude of an object is

$$m = -2.5 \log_{10} \left(\frac{F}{F_{zp}} \right), \tag{2.10}$$

or if the zero point is in magnitudes,

$$m = -2.5 \log_{10}(F) - m_{zp} \tag{2.11}$$

2.5 Filters and Filter Transformations

Optical astronomical cameras are sensitive to radiation from the ultraviolet to the infrared and are unable to distinguish between, for example, a red photon and a blue one. Domestic digital cameras produce colour images, so how is this done? If you take a domestic camera apart and look at the light-sensitive chip, the **charged**

coupled device, or **CCD**, or more likely a complementary metal-oxide semiconductor, or **CMOS**, you will see a coloured grid over the surface of the CCD made of a series of green, red, and blue squares. We will discuss the difference between CMOS and CCD chips in later chapters. Looking closely, you will see twice as many green squares as red and blue. Each coloured square covers a pixel and lets only the light of the colour corresponding to the colour of the square pass through to the pixel. The process of letting only part of the spectrum pass through is known as **filtering**, and the optical device that does this is a **filter**. The camera's firmware joins together four pixels (one blue, one red, and two green) to form one colour pixel. The reason that there is twice as much green is due to the need to make the picture match what you perceive. The human eye is much more sensitive to green than to red and blue, unlike the imaging chip, so twice as many green pixels are used to allow for this.

Looking at the CCDs of most astronomical cameras, we would see no such grid (some exceptions exist, but those cameras are not considered to be of research quality and are aimed at amateur astronomers). This is because we wish to be able to change the range of wavelengths we want to filter out, and we need to control precisely the characteristics of the filter in use. Hence, rather than filter individual pixels, we filter the whole array. Typically, multiple filters are held in a **filter wheel** mounted between the camera and the telescope. Issuing a computer command causes the wheel to rotate so that the correct filter is in the optical pathway. Filters come in a range of sizes and can be either circular or square mounted or unmounted.

Filters are divided into two classes: **broadband**, which are transparent over a wide range of wavelengths, and **narrow band**, which let light from only a specific spectral line pass, allowing for a small degree of movement and broadening of the line. Each filter type has a set standard, so that observations at different observatories using the same filters will produce the same result. Typically, broadband filters will come in a set, with each member assigned a letter that is known as the **band**. Hence the V band (V for visual) covers a region of the spectrum from 511 to 595 nm, and the R band covers the region from 520 to 796 nm. However, you should be aware that filter names are an indication of the range they cover and not their profile with that region. The exact profile is dependent on the photometric system being used. Also, filter names are case sensitive, so a V-band filter is not the same as a v-band filter. Some filter names also have a prime symbol associated, so a V-band filter is not the same as v́, and v is not the same as v́, although since they should cover the same region of the spectrum, they will be broadly similar, but for scientific purposes you cannot use, for example, an R filter when an r filter is needed without employing a complicated filter transformation calculation.

A filter set in wide use in the astronomy community is the Johnson–Morgan UVB and its extended form the Johnson–Cousin set of filters UBRVI (and the very similar Bessel set), which were designed so that a CCD's colour response is similar to that of photographic film. The Johnson–Cousin set consists of a blue (B) filter, a red (R) filter, a green (V) filter, an infrared (I) filter, and an ultraviolet (U) filter. Since almost all imaging is now done with a CCD, the original purpose of the Johnson–Cousin set has been eroded, and its drawbacks, for example the fact that some filters overlap, is making it less attractive for astronomy. However, there is a very large body of work that uses *UBVRI*. Observatories are now moving to Sloan filters, which are known as ú, ǵ, ŕ, í. Sloan filters are an improvement over UBRVI, since they do not overlap and have flatter transmission curves. However, you should be aware that there are multiple "standards," over 200 at last count, with many telescopes, both ground-based and space-based, using variations of, for example, the Johnson–Cousin set. Within this book, this difference should not be a problem, since you should be using the same filter set all the time. It may, however, cause problems if you are using data from elsewhere. A number of good websites show the transmission curves for individual filters, and it is possible to convert from one system to another, especially if the spectral class of the target is known. However, this can be time-consuming and may introduce additional uncertainties into your results.

You might also encounter RGB filters, which shouldn't be confused with Johnson–Cousin R and B filters, since RGB are imaging filters and are non-photometric. In many ways they are the same filters that we see on colour CCDs. For scientific purposes, they should be avoided, since there is no RGB standard, and your results will be difficult to reproduce. The RGB sets are the imaging filters often used by amateur astronomers to produce colour images, although similar results can be achieved using the Johnson–Cousin RVB filters. However, as you might expect, an RGB filter set is considerably less expensive to purchase than an RVB set.

Narrow-band filters tend to have a transmission window only a few nanometers wide, and in some cases, such as a solar H-α, which is much narrower than an astronomical H-α, they may be less than a tenth of a nanometre wide.[1] Many narrow-band filters are centred on a forbidden line, which should be notated by putting the emission line in square brackets, whence OIII should be notated as [OIII], and SII as [SII]. However, H-α is not a forbidden line, although it is a narrow-band filter. As mentioned above, in astrophysics, forbidden lines are emission lines that cannot be produced in the laboratory. The mechanism needed to produce the excited state has a low probability of occurrence and tends to be collisionally suppressed in all but the very lowest densities, which although common in deep space, are all but impossible to achieve in a terrestrial setting. When the [OIII] line was first detected in planetary nebulae, it was thought to be an emission line from a new element, dubbed Nebulium. Forbidden line emission is an important component of cooling of low-density astronomical environments such as planetary nebulae; hence emission line filters are an important tool in observing such objects (Fig. 2.1).

Filters form part of the optical pathway, so changing the filter will change the optical characteristics of that pathway. To avoid refocusing every time an image is taken, a set of filters should have the same refractive index, so that within a set, the same type of glass is used, but it is differently doped to achieve the correct transmission characteristics. A filter set may include a clear, i.e., transparent, filter with the same characteristics as the coloured filters. If a blank space on the filter wheel was used instead of a clear filter, the telescope will have to be refocused in changing to the clear filter and then again on moving away. Thus a physical clear filter is highly desirable.

---

[1] Use only specially designed solar H-α filters to observe the Sun. Standard H-α filters are **unsafe**.

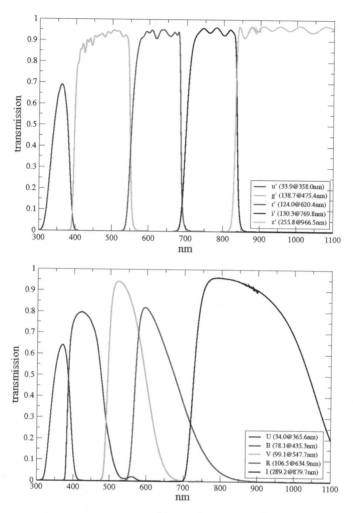

Fig. 2.1 Comparison between the filter profiles of Johnson and Sloan filters. *Source* Leibniz-Institut für Astrophysik Potsdam (AIP)

Other filters may be encountered, which are mostly used for visual observation, but they may also be used in conjunction with a CCD. **Light pollution** filters are designed to suppress emission lines associated with artificial lighting, especially the characteristic sodium discharge lights. **Polarisation filters** may be used to examine the polarity characteristics of a light source, although they are more commonly used in pairs to reduce the amount of light received from a bright object, such as the Moon. This light dimming can also be achieved by the use of a **neutral density filter**, or NDF, which uniformly reduces the amount of light received in observing, for example, the Moon. Other filters exist to aid in the visual observation of emission nebulae (nebula filters) and to enhance planetary features (Tables 2.2 and 2.3).

Table 2.2 Wavelength comparison between Johnson and Sloan filters

Johnson			Sloan		
Filter	Centre (nm)	Width (nm)	Filter	Centre (nm)	Width (nm)
U	365	70	ú	353	63
B	440	100	ǵ	486	153
V	550	90	ŕ	626	134
R	700	220	í	767	133
I	900	240	Other		
J	1250	240	Y	1020	120
H	1650	400	L	3450	724
K	2200	600	M	4750	460

Table 2.3 Wavelength comparison of a sample of narrow-band filters

Filter name	Wavelength (nm)
H-Alpha	656
H-Beta	486
[O-III]	496 and 501
[S-II]	672
He-II	468
He-I	587

On occasion you will have to make a transformation between one filter system and another. There is generally no reason to do this if you are observing in only one band, for example if you are observing an exoplanet transit, but for more complex observations using two or more filters, it is sometimes necessary to make a correction for the filters being used. As stated previously, this can be a complex process that depends not only on the two filter systems but also on the target being observed and its magnitude. Such transformations will introduce inaccurate photometric measurements, and you should be aware that transformed magnitudes may not be reliable. The **Sloan Digital Sky Survey III** (SDSSIII)[2] is a good place to start your filter transformations. It includes filter transformations from Sloan to Johnson for a wide range of target types.

2.6 Colour

For human beings, the perception of colour is a mixture of inputs from the three forms of cones in the retina as well as some complex image processing in the visual cortex of the brain, which is highly sensitive to the language and culture of the individual.

---

[2] Available at http://www.sdss3.org/dr8/algorithms/sdssUBVRITransform.php.

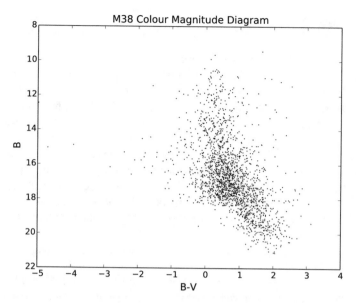

Fig. 2.2 A B-V to B colour magnitude of the open cluster M38 using the USNO Nomad database

This means that in some cases, two people with perfect colour vision will not be able to agree on what colour they are observing. In astronomy, colour has a very specific and very important meaning. It is the ratio of the observed fluxes in two filters (or magnitude A minus magnitude B).

As discussed earlier, apparent magnitudes are dependent on the distance from the observer, but the colour is distance independent but dependent on the spectral class. Hence plotting a colour, say V-B, against a magnitude, say B, to produce a colour–magnitude diagram (CMD) provides important information about stellar mass and stellar evolution, In fact, such a plot was undertaken by Ejnar Hertzsprung and Henry Norris Russell in 1910. The **Hertzsprung–Russell (HR) diagram** is one of the most important plots in astrophysics. It enables astrophysicists to identify the key stage of stellar evolution, the **main sequence**, a dense band of stars running from top left to bottom right in which stars are undergoing core hydrogen fusion.

This observation coupled with the observation that as stars evolve they turn off the main sequence enables astronomers to age star clusters using the colour–magnitude diagram by looking for the main sequence turnoff point in the HR diagram.

Figure 2.2 shows a B-V versus B colour–magnitude diagram of the open cluster M38. This plot contains a large number of stars that are either foreground or background objects and are therefore not part of the cluster. Likewise, due to the large distance to M38, interstellar dust has preferentially scattered blue light over red light (a process known as **reddening**), resulting in an increase in the B-V colour.[3]

---

[3] Recall that low magnitudes are increasingly bright, so if there is less light at the blue end, the blue magnitude increases in value, and hence so does the B-V colour.

2.7   Polarisation

As stated previously, light is an electromagnetic wave consisting of an electric field perpendicular to a magnetic field, both of which are perpendicular to the direction of travel. In general, the orientation of the electric field is random. However, certain physical processes can cause the electric field to be inclined at a preferential angle to the direction of movement. If this is the case, the light is said to be **linearly polarised**. Additionally, the electric and magnetic fields may rotate, with their axis of rotation the direction of travel. Hence in this case, the light appears to draw out a helix. We can characterise a wave as being either left-handed or right-handed, depending on the direction of rotation seen from the position of the observer. Again, in general, the distribution of the fields' handedness is random, but physical processes can cause light to exhibit a preferential left- or right-hand rotation. In this case, it is said to be **circularly polarised**.

Polarisation plays an important part in astronomy, with linear polarisation being significant in the study of interstellar dust and planetary atmospheres, whilst circular polarisation is a key investigative tool in understanding the nature of very large scale magnetic fields, especially in radio astronomy. In general, besides observations of the Sun, work with polarisation is beyond the capability of small telescopes, since the degree of polarisation from a source can be very small (Fig. 2.3).

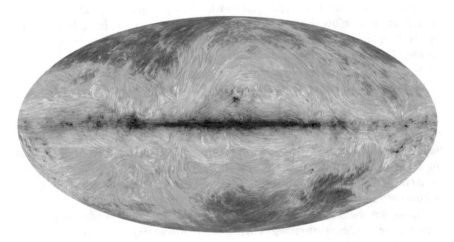

Fig. 2.3 Magnetic field lines traced by dust emission at 353 GHz by the Planck Space Telescope. Image European Space Agency 2015

Chapter 3
The Telescope

Abstract The telescope is the principal tool of the astronomer. Therefore, it is important to understand the type of telescope, the process by which the telescope is mounted, its limitations, and the accessories available to the astronomer. Furthermore, since on occasion, telescopes suffer from optical problems, we show how to identify such problems and to resolve them if possible.

3.1 Telescopes

Before beginning this chapter, we need to issue you a small but necessary warning. Under no circumstances should you look at the Sun with any optical instrument or with the naked eye unless the observation is by projection or with a special solar filter. If you use a filter, I strongly suggest that you put a camera on to check that the filter is not damaged before you put your eye to it. You can easily replace a camera, but replacing your eye is not so easy. To illustrate the damage, some years ago, the author was using a solar telescope mounted on a much larger telescope. Of course, I put the lens cap on the main telescope, as pointing it at the Sun without the lens cap would destroy the very costly astronomical camera mounted on it. Unfortunately, I forgot about the small 5 cm finder scope mounted next to (and hidden by) the solar telescope. In the two seconds between pointing towards the Sun and realising that the lens cap was off and replacing it, the steel crosshairs in the finder scope melted. Fortunately, finder scopes are cheap, but it was a lesson learned. So take heed, and always double check!

A telescope has two functions, to gather light and to bring that light to a focus. These manipulations may be achieved in two ways; we can refract the light by means of lenses or employ mirrors to reflect it. You will typically be using a derivative of a reflecting telescope, although your finder telescope or guide telescope, if you have one, will normally be a refracting telescope. Likewise, if you use a solar telescope, it is likely to be a refracting telescope.

Independent of its type, the telescope will have a number of important optical characteristics that you should be aware of:

© Springer Nature Switzerland AG 2020
M. Gallaway, *An Introduction to Observational Astrophysics*,
Undergraduate Lecture Notes in Physics,
https://doi.org/10.1007/978-3-030-43551-6_3

- The **aperture** size is the diameter of the telescope and is a measure of its light-gathering power. Several techniques in telescope design may result in the optical aperture being, in effect, smaller than the physical aperture.
- The **focal length** (f) is the distance separating the first optical component and the first point of focus.
- The **field of view** (FOV) is the amount of sky visible through the instrument, measured in degrees or arc minutes. The angular field of view of a telescope (in radians) is the aperture of the eyepiece (the field stop) or the diameter of the detector f_o divided by the focal length f:

$$fov = \frac{f_o}{f} \times 206{,}265 \qquad (3.1)$$

- **Magnification** is the ratio between the angular size of an object observed through the telescope its angular size when it is observed without the telescope. For unresolved objects such as stars, the magnification has little or no effect. For smaller consumer telescopes, high magnifications are often quoted. In practice, high magnifications are rarely used in astronomy and generally cause more problems than they are worth. If you are using an eyepiece, the magnification is simply the ratio of the focal length of the telescope, f, and the focal length of the eyepiece, f_e; if, however, you have an astronomical camera at the focus, then the nearest approximation to magnification is the image or plate scale. This is merely the focal length of the telescope, f, over the diagonal diameter of the light-sensitive chip in the camera (the CCD):

$$Mag = \frac{f_e}{f} \quad Plate\ Scale = \frac{f}{f_o}. \qquad (3.2)$$

- Another often quoted feature of telescopes is the f number. This is simply the focal length f divided by the optical aperture size, A. In general, telescope with low f numbers are called *fast*, because they gather light quickly due to their large field of view. However, the downside is that they have reduced resolution. Additionally, optics for fast telescopes tend to be very expensive to manufacture in comparison to those with high f numbers:

$$F = \frac{f}{A}. \qquad (3.3)$$

- The **angular resolution** of the telescope (α_c) is the telescope's ability to separate two objects; it is given by the average wavelength of the observed spectrum divided by the aperture of the telescope D_0. This is the diffraction limit caused by the formation of the Airy disc, which is produced by the diffraction of light within the optics:

$$\alpha_c'' = \frac{1.22\lambda}{D_o} \times 206{,}256. \qquad (3.4)$$

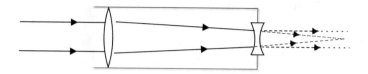

Fig. 3.1 Diagram showing the optical pathway of a Galilean telescope

Although in attempting to resolve two stars, the Rayleigh criterion (3.5) (which gives the resolution in radians) is more representative, the simpler Dawes' limit is also widely used (D is in millimetres). We have

$$\alpha = \sin^{-1} \frac{1.22\lambda}{D} \qquad (3.5)$$

and

$$\alpha \cong \frac{115.82}{D}. \qquad (3.6)$$

In reality, most telescopes are not diffraction limited but are limited by atmospheric turbulence, that is, **seeing limited**.

The first refracting telescope was made in the early sixteen hundreds in the Netherlands by spectacle makers. It was not until Galileo Galilei designed and constructed his own a telescope a few years after their first appearance that the telescope began to be used for astronomy. Galileo's telescope used a convergent objective lens and a divergent eyepiece, which, unlike modern instruments, produced an upright image.[1] Galileo's design was rapidly improved upon by Johannes Kepler, who changed the divergent eyepiece to a convex one and thereby inverted the image (Fig. 3.1).

The Keplerian telescope allowed the use of increased magnification but failed to overcome two of the refracting telescope's principal problems, the first of which is an engineering problem, due to the fact that the mass of an objective lens increases disproportionately with its diameter. Not only does this make the telescope unwieldy to use, but the lens becomes difficult to mount due to the thinness of the glass at the edge, and it tends to sag under gravity, resulting in altered optical characteristics with elevation. Additionally, as the glass volume increases, it becomes increasingly more difficult to keep the lens free of imperfections (Fig. 3.2).

You will recall that a lens's refractive index is wavelength specific. This being the case, a lens will bring each colour to a different focus, resulting in coloured rings around objects, a phenomenon known as **chromatic aberration**. This can be overcome in part by using much longer focal lengths. However, it was not until the introduction of **Achromatic**, and later **Apochromatic**, lenses that the problem was resolved. Achromatic lenses use two types of glass with different refractive

---

[1] Most modern telescopes invert the image.

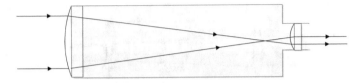

Fig. 3.2 Diagram showing the optical pathway of a Keplerian telescope

indexes bonded together to form a lens with the same refractive index for red and blue light. Apochromatic lenses use the same concept but have three components, for red, blue, and green. The problem with mass has never really been resolved, and by the beginning of the twentieth century, refracting telescopes reached (and possibly even exceeded) their maximum operational size. Over the years, additional optical coating has been applied to astronomical lenses, and although this has improved them, it has not overcome their physical limitations. Of course, telescopes must be exposed to the elements, although they should not be made to encounter very extreme weather, and over time, a lens becomes pitted, a coating becomes worn, and the optical bonding compound used to bond achromatic and apochromatic lenses loses its optical transparency, at which point the lens will either need maintenance or replacement.

The concept of using mirrors to magnify an image had been known for at least five hundred years before their successful application to astronomy by Newton. Mirrors have a number of advantages over lenses; only the surface of the mirror needs to be optically smooth, reducing the need to produce optically perfect glass; mirror-based telescopes do not suffer from chromatic aberration; and the mirror's position at the base of the telescope makes it easier to operate. However, mirrors are not entirely without design problems. Typically, a secondary mirror is needed near the primary focus, which then partially obstructs the optical path and reduces the efficiency of the telescope. Also, light is diffracted around the secondary mirror's supports and the secondary mirror itself, which causes the spikes and halos seen around some stars in astronomical images, although admittedly in some cases such artifacts might have been added digitally for effect. There is also a tendency for light incident on the edge of the mirror to be brought to a slightly different focus from that of light at the centre. As a result, objects become smeared out, a problem known as **coma**. An additional problem is the alignment of the primary and the secondary mirrors. If this alignment is not within a very small tolerance (Collimated), the light arrives at the focus at a slight angle to the focal plane, and consequently, point sources become elongated.[2] **Collimation** problems normally worsen over time, but they are easy to address. Hence recollimation is part of the normal maintenance procedure for a telescope.

As with lenses, mirrors become increasingly massive as their diameters increase. Sagging under gravity is prevented by supporting the mirror across its entire rear surface area. However, it appears that the engineering limit for producing astronomical-

---

[2]They are often described as looking comet-like.

grade mirrors has been reached at around 8 m. A number of 10 m class telescopes exist beyond this limit, but those have segmented mirrors, which are manufactured as individual components and then combined using a technique that is not itself without problems. The planned 30 m+ class telescopes currently under construction, including the appropriately named 39 m European Extremely Large Telescope (E-ELT), all have segmented mirrors.

A wide range of designs for reflecting telescopes exist, most of which differ only in the path the light travels. The simplest and perhaps the least useful is the prime focus telescope. This has no secondary mirror. Rather, the detector is placed at the prime focus. Although of little use for observing with the eye, the introduction of the camera to astronomy, and in particular the digital camera, has made this design more practicable. In fact, many parabolic dish radio telescopes are of this design.

Newton's original design is still in wide use around the world, particularly with amateur astronomers due to its low cost and ease of operation (Fig. 3.3). The Newtonian telescope uses a flat mirror set at 45° to the primary mirror and just inside the focal point, so that the point of focus is outside the tube, often close to the top. This allows for the easy mounting of a focuser at what is normally a comfortable position for the observer. However, in mounting an instrument, the focal point of a Newtonian telescope is problematic, as the additional mass of the instrument so far up the tube makes balancing difficult, although not impossible, and for very large telescopes, the focal point becomes increasingly inaccessible. Consequently, most professional instruments have moved to a Cassegrain-like design, although a small number of operational instruments retain the ability to switch between Newtonian and Cassegrain operation.

The Cassegrain telescope uses a hyperbolic secondary mirror placed parallel to the parabolic primary mirror, which reflects the light down the tube and through a hole in the centre of the primary with the focal plane lying just outside of the tube at its rear. Hence, Cassegrain telescopes are viewed through the bottom, very much like a traditional refractor, which allows the positioning of the instrument in a more convenient location. This also means that a Cassegrain telescope tends to be physically shorter than a Newtonian of the same focal length. By modifying the Cassegrain by replacing the primary with a hyperbolic mirror, suitably matched to

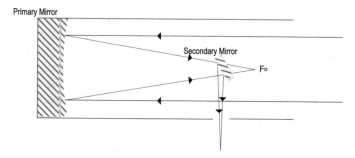

Fig. 3.3 Diagram showing the optical pathway of a Newtonian telescope

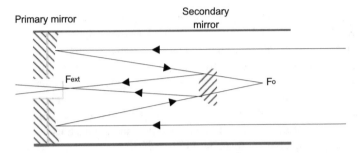

Fig. 3.4   Diagram showing the optical pathway of a Cassegrain telescope

the secondary, George Willis Ritchey and Henri Chrétien introduced a nearly coma-free and spherical-aberration-free telescope design that now bears their name and perhaps, more importantly, is now the most commonly used professional telescope design. Examples include the four 8.2 m Very Large Telescopes in Chile, the twin 10 m Keck telescopes indexKeck in Hawaii, the Hubble Space Telescope, and the Spitzer Space Telescope (Fig. 3.4).

The telescope you will be using as an undergraduate may well be neither a reflector nor a refractor but rather a hybrid of the two, a **catadioptric** telescope. Catadioptric telescopes use a correcting lens (known as the correcting plate) in front of the primary mirror to counter optical imperfections and reduce the physical focal length of the telescope whilst retaining a long actual focal length. This also allows the use of a spherical primary and secondary. Spherical mirrors are much easier to manufacture, and as a consequence are much cheaper. Most catadioptric telescopes have a full-diameter correcting plate and are known as **Schmidt Cassegrain** telescopes, or SCTs. The positioning of a full-width correcting plate at the front of the instrument allowed the introduction of a further simplification, the **Maksutov-Cassegrain** telescope, in which the secondary is actually the curved mirrored surface of the correcting plate. This design allows for low-cost construction and maintenance, as the secondary is securely fixed and the tube sealed, reducing wear on the optical surfaces. The low manufacturing cost of these designs, their intrinsically low maintenance, and the fact that they are easy to handle due to their compact design have made these telescopes very popular with educational facilities and serious amateurs (Fig. 3.5).

Most professional research-grade telescopes are of the Ritchey–Chrétien design, itself a version of the Cassegrain. The Ritchey–Chrétien uses a matched primary-secondary non-spherical pair to produce a telescope that is compact for its focal length, has a wide field of view, and is largely free of coma, although not of astigmatism. However, astigmatism in this case can be moderated by the introduction of either a third mirror into the optical pathway or the use of corrective lenses.

Here is an important point about the condition of optics. If you think that a telescope has dirty optics or is suffering from condensation, never wipe the optics clean. In fact, never clear them yourself at all. If you wipe them, small dust particles will scratch the optical surfaces. Your observatory staff will know how to clean the

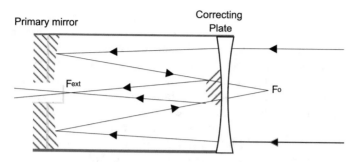

Fig. 3.5 Diagram showing the optical pathway of a Schmidt-Cassegrain telescope

optics and will have specialised tools and chemicals to do so; hence you should leave it to them. If you are using an SCT and are condensating, you might find that slowly warming the correcting plate with a hair dryer solves the problem, but do not hold the hair dryer too close. If that is ineffective, the humidity is likely too high for good observation, in which case it is normally worthwhile shutting the telescope down until conditions improve.

As with refracting telescopes, reflecting telescopes suffer from optical degradation due to weather (in fact, due to just being exposed to the air). With time, the thin silver or aluminium coating on the mirror degrades and will eventually have to be replaced—re-silvering. This is a relatively easy process (depending on the size, and replacing the mirror is not as easy as taking it out), but it requires a vacuum chamber, although many large professional observatories have silvering machinery on site.

3.2 Mounts

Just as important as the optics is the structure to which the telescope is fixed that allows it to rotate. If the structure is not stable enough, finding an object and keeping it in the field of view becomes extremely difficult and is perhaps the biggest problem with the most inexpensive telescopes, which are good only for the Moon and the brighter planets. A wide array of mounts are currently in use including Dobsonian (which tends to be shortened to just Dob), German equatorials, GoTo mounts, English equatorials, and Yoke mounts, although effectively only two designs are in use.

The first mounts were of an altazimuth (alt-az) design, whereby the telescope is moved around both the vertical (altitude) and horizontal (azimuth) axes. This design is cost-effective and easy to use. It does, however, have a major drawback. Astronomical objects appear to move across the sky in an arc as a result of the Earth's rotation and its inclination to the plane of the solar system (more on this later). The effect of this is that in order to track an object with an alt-az mount, the mount needs to drive both axes in a nonlinear fashion. If it does not, the object soon drifts out of the field of view. Such tracking is challenging to accomplish by hand and is extremely difficult to do mechanically without the use of a computer. A camera fixed to a telescope using an alt-az mount rotates as the mount tracks, which can cause problems with longer exposures. There is a simple solution, however: the equatorial mount. If you tilt the mount so that the centre of rotation of the horizontal axis now points to the celestial pole and the vertical axis is perpendicular to the inclined horizontal axis, the telescope naturally follows the same arc as stars, and observed objects do not drift in the direction of the other axis. This orientation is known as polar alignment, which means that we need to drive only a single axis, which can be done simply with clockwork. Multiple designs of equatorial mounts exist, one of the more popular being the German equatorial mount (Fig. 3.6).

Equatorial mounts tend to be heavy, due to the amount of required counterweighting, and they need a larger dome to house them than do alt-az telescopes. They also suffer from what is known as **meridian flip**. Meridian flip occurs, not unsurprisingly, when the telescope transits the meridian. In order to stop the telescope from hitting the mount, the telescope must be reorientated, or flipped. This means that if you are performing a series of observations of an object, there will be a break as the object crosses the median, which is the very best place to acquire an image.

With the introduction of small, cheap, powerful computers and field rotators that rotate the camera so that it maintains the same position relative to the field, the

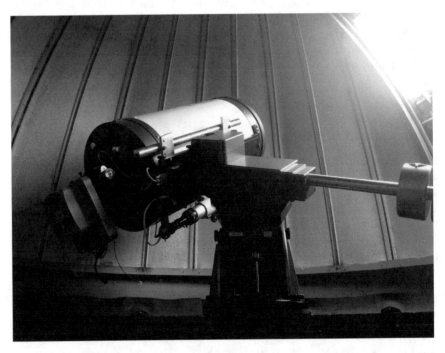

Fig. 3.6 Image of a telescope mounted on a German equatorial mount. Image courtesy the University of Hertfordshire

limitations of the alt-az mount have now been overcome. Consequently, almost all modern large telescopes are of the alt-az design, although equatorial mounts are still common.

Pioneered by amateur astronomer John Dobson, the Dobsonian is one of the simplest astronomical telescopes. In simple terms, it is optically a Newtonian, mounted in an open boxlike structure, supported by two pegs that extend from the tube and pass through the box. Because the bulk of the mass of a Newtonian is in the primary mirror, the centre of gravity tends to be located near the primary—the feature that makes the Dobsonian possible. The box, which provides the alt-az movement, is mounted on a flat plate with a single fixing passing through the base of the box at its centre. Typically, there is a low-friction bearing (often Teflon) between the two surfaces. Since its invention, the Dobsonian has evolved into a wide range of differing, although fundamentally similar, designs. For example, thee are now open tube Dobsonian and push-go Dobsonian, which have encoders linked to a tablet PC or smartphone that allows unpowered GoTo (the user pushes the tube to the target under the direction of the tablet or smartphone), and there are even powered tracking GoTo Dobsonians. The big advantage of Dobsonians over other telescopes is their price. In most cases, a good mount costs as much as the telescope to which it is mounted. The very low cost (many are still built in home workshops) of this mount has resulted in an increase in aperture size, giving amateur astronomers access to deep sky obsessing that a few decades ago would have been out of reach for many. However, for astrophotography, the Dobsonian struggles to compete with more sophisticated mounts such as the German equatorial. However, there is a lot to be said for having what effectively is a huge light bucket for observing the night sky, especially one as light and portable as the Dobsonian. Australian amateur astronomer and minister Robert Evans discovered a large number of extragalactic supernovae from his backyard using his 31 cm Dobsonian. This can be attributed mostly to his ability to identify supernovae by eye but also to the ease of use of the Dobsonian design and to the good weather and clear skies of Australia.

Perhaps one of the least common telescope mounts that you are still likely to encounter is the transit telescope. Transit telescopes point only at the meridian (the line that runs north to south) and consequently can move only in elevation. Transit telescopes, which tend to have very long focal lengths, were used to determine when an object crossed the local meridian, and its elevation, which enables the observer to determine both the local time (and hence their position on Earth) and the location of the object in the sky. Transit telescopes tend to be found in older observatories, and new constructions are rare. The monument to the Great Fire of London, near Pudding Lane, contained, somewhat bizarrely, a transit telescope.

One of the biggest changes in telescope mounts in recent decades is the introduction of GoTo telescopes, and more recently robotic telescopes. GoTo telescopes feature powered mounts with encoders that feed their orientation back to a computer. If the telescope is portable, it may feature a Global Positioning System (GPS) receiver to aid in the setup. Effectively, a GoTo telescope always knows where it is pointing, although some degree of setup is needed, which normally consists in manually pointing the telescope at three bright stars so the computer can work out the translation from encoder position to actual sky position. Once a GoTo telescope is set up for the night, finding a target is just a case of picking a target from a list held on the computer or telescope handset and pressing a button, although you may occasionally need to re-align the telescope during the night, as position errors can creep in. Gone are the days of finding a bright star first and then using a star map to hop between stars to find a faint target.

3.3 Eyepiece

It is likely that for most of your observation time, you will not be looking through the telescope. This is especially true for remote and robotic observations. However, sometimes you do need to use your eyes, for example in resolving a problem or for outreach and public engagement events at your observatory.

You cannot just look through a telescope. You need additional optics to bring the image to focus on your retina, namely an eyepiece.

Eyepieces come in a bewildering array of types, but all do the same thing. They take the light from the telescope and bring it to focus for an observer, although occasionally you might need an eyepiece for an instrument if it cannot operate near the primary focus. In general, eyepieces come in a range of barrel sizes. The barrow is the part that slides into your focuser and is typically chromed. Standard sizes are 1.24 and 2 in., although 1 in. and sizes greater than 2 in. exist. Not that the size listed is the size of the barrel, since many top-end eyepieces are wider than this in their main body (this is to fit in all the optics). Adapters are available if your eyepiece is too large or too small for the focuser barrel.

On the eyepiece, there should be a number, normally in millimetres, noting the focal length of the eyepiece, along with with a type and perhaps a manufacturer's name. We will ignore the manufacturer's name, as there are many, and it is largely unimportant.

The magnification of the telescope with the eyepiece is the focal length of the instrument divided by the focal length of the eyepiece. Unless you are looking at a planet or the Moon, it is likely, at least initially, that you will not want a high magnification, as that limits the field of view.

The point at which you have to place your eye to get the full field of view when using an eyepiece is known as the eye relief. The farther from the eyepiece, the better, with at least 15 mm needed for those wearing glasses. Short eye relief eyepieces are uncomfortable to use, especially if they are so short that the user's eyelashes touch the lens.

The exit pupil is the size of the beam of focused light coming out of the eyepiece. The exit pupil is the focal length of the eyepiece divided by the telescope's focal ratio. Hence a 30 mm eyepiece used with a 2000 mm focal length telescope with an aperture of 200 mm will have an exit pupil of 3 mm. You should try to get an exit pupil size of between 5 and 7 mm. Human dilated pupil diameter shrinks with age, so bear this in mind.

The choice of eyepiece will affect your field of view, although that is also, to a large degree, dependent on the telescope. Some eyepieces are specially designed to give a larger, more desirable, field of view than others, although this comes at a financial cost.

Telephoto (also known as zoom) eyepieces are becoming increasingly popular amongst amateurs. These eyepieces have variable focal lengths, for example, 8–24 mm. They enable an observer to zoom in and out without changing the eyepiece. However, keep in mind that every eyepiece design is a compromise between focal length, optical quality, weight, the field of view, eye relief, and cost, and furthermore, a zoom eyepiece is never going to compete for accuracy with a fixed focal length lens with a comparable design.

The simplest eyepiece consists of just a single concave or negative lens. The Galilean eyepiece is a low-cost design giving a very poor field of view. They are best used in low magnification scenarios such as toy telescopes.

Like the Galilean, the Keplerian eyepiece consists of a single lens, but in this case it is a convex lens. It has a wider field of view than the Galilean.

The Huygens eyepiece was the first eyepiece to use multiple lenses. It consists of two planoconvex lenses facing the same direction, in series, with an air gap. It suffers from short eye relief, high image distortion, chromatic aberration, and a poor field of view. However, its low cost means that it is often shipped as the standard lens for bottom-end telescopes. Since its components are separated rather than bonded, it is not subject to damage by heat from the Sun; hence Huygens eyepieces are often used for solar projection.

A slight improvement on the Huygens eyepiece was achieved by the use of two planoconvex eyepieces placed one focal length apart with both convex sides facing inwards. The Ramsden eyepiece, like the Huygens, can suffer from chromatic aberration and has poor eye relief.

Eyepiece chromatic aberration was addressed by the introduction of the Kellner eyepiece . The Kellner consists of three lenses: a planoconvex in front of a bonded doublet of a convex lens and a planoconcave lens. The Kellner offers a significant reduction in chromatic aberration compared to earlier designs as well as good eye relief and apparent field of view. It is often shipped with smaller telescopes, and it is quite possible that one is to be found in your observatory.

The Plössl eyepiece is a compound eyepiece consisting of two bonded convex concave-plano doublets. The Plössl has a good field of view but relatively poor eye relief, and it is well suited for deep sky and planetary viewing. It is one of the most common eyepieces in use, and there will almost certainly be one in your observatory. For good performance, the bonded doublet must be well matched to prevent internal reflections, although this is often mitigated by the use of coatings. As a result, there is a significant difference between the top-end and bottom-end Plössl eyepieces.

The orthoscopic eyepiece is a high field of view, low distortion eyepiece, ideal for observing the Moon and other solar system objects. The orthoscopic consists of two doublet convex lenses bonded to either side of a concave lens, with the bonded triplet sitting in front of a planoconvex lens.

Fig. 3.7 An array of eyepieces, with both 1 and 1.5 in. barrels

The monocentric eyepiece is an achromatic eyepiece with a small field of view but good optical qualities such as image brightness and a lack of internal reflections. It consists of two concave-convex crown glass lenses bonded to either side of a flint glass block. Uncommon in teaching observatories, it is popular with amateurs for planetary imaging.

Having a wide field of view and good eye relief, the Erfle eyepiece is a five-component eyepiece best suited for long focal lengths of 20 mm and above. At smaller values, the Erfle tends to suffer from astigmatism and ghosting. The Erfle consists of a bonded convex-concave doublet and a convex concave-plano doublet sandwiching a convex lens.

At the top end of current eyepiece design, and cost, is the Nagler. In its basic form, it has a negative field doublet followed by a series of positive doublets and singles. The Nagler has an excellent field of view and very good eye relief. However, it is complex and can be heavy and costly. The Nagler is very desirable, but given that most telescopes you use will rarely be actually looked through, it is unlikely to be found in your observatory (Fig. 3.7).

You may wish to undertake an observation that is not achievable with the current optical setup, requiring the adjustment of the optical pathway for physical reasons, or to correct an optical defect. The **diagonal** or **star diagonal** changes the orientation of the optical pathway by 90°.[3] This is most often done to accommodate an instrument or to make visual observations more comfortable. The change in the pathway is

---

[3] 45° diagonals are also available.

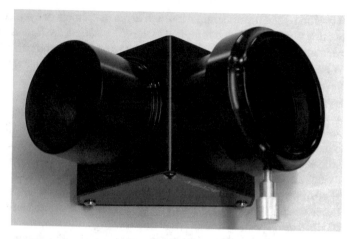

Fig. 3.8 A star diagonal with a two-inch barrel

performed by the use of either a mirror inclined at 45° or a prism. Mirror-based diagonals have the additional advantage that they right the inverted image, although the mirror will deteriorate, and therefore they have a more limited life-span than prism-based diagonals. Modern prism diagonals now allow the transmission of more light than mirror-based diagonals. They can introduce chromatic aberration, but in general, this does not occur when they are used for long focal length instruments such as the SCT, which is mostly used by educational facilities (Fig. 3.8).

On some occasions, the object you wish to observe is resolvable, but the image size is very small. When using a telescope visually, you can generally use a shorter focal length eyepiece, although this isn't always a comfortable choice and not a solution when a camera is used. To overcome this problem, you may try to use a **Barlow** lens. A Barlow is an optical device that sits between the camera or eyepiece and the prime focus, effectively increasing the focal length of the telescope. However, be aware that it might introduce aberration, increase exposure times, and, as it extends the camera out further, balancing problems. A Barlow lens will typically double the focal length, thereby doubling the image size.

Alternatively, you may have a target whose angular diameter is greater than that of your telescope's field of view. In this case, you could take several images and montage them together. However, if this is not possible, you can use a **focal reducer**. This is somewhat like a Barlow and suffers from many of the same problems, but unlike a Barlow, the focal reducer, as the name suggests, reduces the effective focal length of the telescope. Recalling that $fov = \frac{f_e}{f} \times 206,265$, we can see that a focal reducer effectively increases your field of view.

The last item on this, in no way comprehensive, list of items you may encounter between the camera/eye and the focuser is the **flip mirror**. The flip mirror is, as its name suggests, a flat mirror which can be moved from a position where the mirror is parallel to the telescope to one where it is inclined 45° to the telescope. When it is in the parallel position, light travels straight through to the instrument or eyepiece in the normal mounting position. When it is flipped to the 45° position, the light is diverted to another detector at a 90° angle to the primary detector. Hence you can have two instruments attached to the telescope at the same time. Typically, this would be a spectrograph and eyepiece, but not always.

On occasion, you might find it impossible to focus on the image, either visually or with an instrument. This is often caused by the focuser lacking the degree of travel to put the image in focus. This is overcome by use of an extension tube. The tube contains no optics and just allows the positioning of the instrument at a more distant point from the focus.

Parfocal rings are small components that can be affixed to an eyepiece or a camera to match the position of the focuser between two items, thereby removing the need to refocus between swaps. Typically, the matching is between a camera and an eyepiece, and parfocal rings are especially useful when one is using a flip mirror.

If you are using a digital single lens reflex camera (DLSR) for imaging or a webcam, it might be necessary to have a particular adapter. There are known as T-rings and are specific to a camera manufacturer and sometimes to a camera generation. It is not obvious by sight which T-Ring is the right one, so check with your observatory staff and make sure it fits before planning the night's observing.

3.4 Problems

A number of aberrations exist that appear in telescope optics. Some are inherently unavoidable due to the design of the optics, some by the incorrect calibration of the optics. Still others are caused by external factors or by poor operation of the telescope and instrument package.

One of the most common problems I find with students is that their images frequently come out blank. Often, it is just a case of the exposure time of the camera being set too short, the camera having lost connection with the control computer, or the camera shutter being set too close. Most commonly, though, the reason is that the lens cap is still on the telescope! I have seen this happen dozens of times with undergraduate and postgraduate students and even academic staff. If you are getting black images, run through the following checklist before getting help.

1. Is the lens cap on the telescope?
2. Is the slit of the dome (if it has one) aligned with the telescope. If not and you have a tracking dome, park both the dome and telescope and reactivate the startup procedure. Sometimes, the control software gets confused as to where the telescope is pointing and, on occasion, where the dome is. If you have a non-tracking dome, just move the slit to the correct position.
3. Is the exposure time too short? In any part of the sky, a telescope of reasonable size should see at least some stars, but not if the exposure time on the camera is very short. If the previous user was observing something very bright, such as Jupiter, for

instance, the exposure time is likely to have been very short. Increase the exposure time until you can see something other than noise.

4. Display software can stretch an image or display only a certain range of pixel values. This is a very useful feature, but if you are looking at a completely different object from those observed previously, it may result in nothing being displayed. Check the settings.

5. At this point, there might be a problem with the camera or the control software or even a cable. Check whether there is power to the camera and that all the cables are seated correctly, as it is easy for them to be pulled out accidentally without being seen in the dark of an observatory dome. If everything is OK, you will likely have to power cycle the camera and the control software. Unless you have experience doing this, talk to the observatory support staff first.

3.4.1 Optical Problems

When in focus, a star, which is a point source, should have a Gaussian, or bell-curve-like, profile in all directions, as the light from the star should be distributed circularly around the central pixel. If it is not, then more than likely there is a tracking error. However, it might be an indication of something more serious that needs addressing.

We have already discussed chromatic aberration, the appearance of coloured halos around sources, caused by refraction bringing different wavelengths to different focal points. As in all likelihood you will be using a reflector, it is unlikely you will encounter this unless you are using poor eyepieces or you are using a low cost refractor, such as might be the case with your telescope's finder scope. Chromatic aberration manifests as a brightly coloured halo, most often blue, around objects. If your optical system suddenly develops chromatic aberration, you could be experiencing condensation on the optics or be using a poor eyepiece. This is illustrated in Fig. 3.9.

In telescopes that use parabolic mirrors such as Newtonians, an effect known as **coma** can sometimes be seen. Coma is a form of optical aberration which makes point-like sources, such as stars, appear comet-like, in that they appear to have a tail, hence the name. Coma is caused by the fact that parabolic mirrors bring only parallel rays to focus at the same point, and coma may be more pronounced away from the centre of the field. Precise collimation of the optics may help reduce coma, as may the introduction of additional optics. However, in general, coma is caused by the design of the optics. Telescopes using spherical mirrors, correcting optics, or both, such as Schmidt and Ritchey–Chrétien telescopes, suffer from little or no coma (Fig. 3.10).

All optical systems use curved surfaces to bend and focus light. This is not a problem for the human eye, as the retina itself is a curved surface, but for a camera, the projection of a curved image onto a flat surface can cause a problem known as **field curvature**. The result of field curvature is that objects near the edge of the

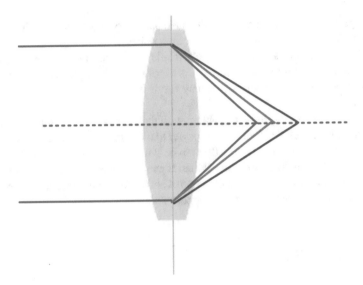

Fig. 3.9 Diagram showing chromatic aberration. Note how red, green, and blue light splits to arrive at different focal points

Fig. 3.10 Diagram showing coma in a refracting telescope, but the same effect applies to reflecting telescopes

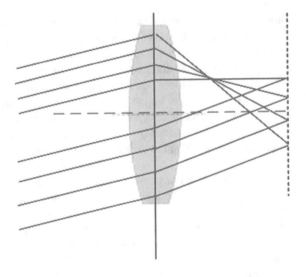

frame may appear out of focus. This is generally not a problem for cameras with small chips, but it becomes increasingly problematic for those with large ones. This effect is illustrated in Fig. 3.11.

The sudden onset of coma could be a tracking error, especially for long exposure images. If the mount is tracking too fast, too slow, or not at all, then the stars will appear to trail and become elongated. If you suspect a tracking error, park the telescope and go through the startup procedure again. If the coma persists and it

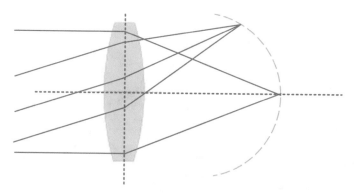

Fig. 3.11 Diagram showing field curvature. Note how light near the edge of the field is brought into focus in front of the focal plane formed by the centre of the field

isn't a tracking problem, then it's likely to be an optical alignment issue, which isn't something that is easily fixed and should be left to your observatory support staff.

Spherical aberration, which is caused by different parts of the mirror or lens bringing light to focus at different points, causing rings to appear around stars, is infamous for being the optical error that affected the Hubble Space Telescope (HST) shortly after launch. As this is purely a function of the optics, it must be overcome by the design of the optical system. The easiest method is to use a matched secondary mirror. Alternatively, correcting optics, such as the corrector plate in a Schmidt, can adjust the optical pathway so that there is no spherical aberration. This was the initial solution that saved the HST, with the COSTAR optics package being placed in the optical pathway, replacing the high-speed photometer. Later, as the HST's cameras were replaced, correctors were built into the instruments, and in 1993, COSTAR was removed and replaced by a new instrument (Fig. 3.12).

Vignetting is a common problem in optical systems and is to some degree unavoidable. Vignetting is often characterised by the centre of the field being bright and the outer edges less so. The result is a bright circle often in the centre of the field of view. Vignetting is caused by obstructions in the optical pathway, including the tube, the secondary, any structure holding the secondary, the hole in a primary Cassegrain mirror, and any anti-reflection baffles. When you are using a digital imager, vignetting can be addressed in the calibration process.

Consider the intersection of two perpendicular planes coaxial with the light path. If the two planes are not correctly aligned, there is a tendency for one plane to extend the image along one axis. Instead of getting a nice round star, we get a star with an elliptical or lenticular shape. This effect is known as **astigmatism**. At first, it may seem similar to coma. In astigmatism, however, the light is spread equally along the axis, whilst with coma it is not. Hence for most scientific purposes, astigmatism is preferable to coma. In some optical systems, such as the Ritchey–Chrétien telescope, a tradeoff is made between reduced coma and increased astigmatism. In general, most

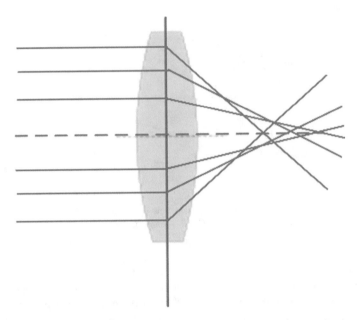

Fig. 3.12 Spherical aberration diagram showing how images near the edge of the system are brought to a different focus from those near the center

telescope designs do not suffer from astigmatism, and those that do can use additional optics to correct it, as happens in some Ritchey–Chrétien telescopes.

A less common effect is **pinched optics** (in fact, the author has never encountered it outside the lab). It is an effect that makes stars appear slightly triangular. It is caused, as is suggested by the name, by optics mounts that a pinching, and therefore deforming the optics. Normally, this would be encountered during first light for a telescope or when the telescope has been rebuilt, for example when the mirror has been resilvered or a lens doublet rebonded, and is being tested. It might also be encountered in extremely hot or cold weather, when expansion or contraction of the optics or their mounts causes the pinching. If you encounter pinched optics outside a test observation, you should be asking yourself why are you observing in those conditions at all.

In order to produce the best possible optical image, the optical components must be correctly aligned, or **collimated**. Otherwise, the image will be distorted. Over time and with use, most telescopes slowly lose collimation and have to be re-collimated. The easiest way to determine whether a telescope is losing collimation is to look at an-out-of-focus star. It should appear doughnut shaped, with the dark disk making the hole in the doughnut, this hole being the shadow of the secondary. The secondary should appear exactly in the centre of the out-of-focus star. If it does not, then the telescope is out of collimation. Newtonian telescopes, and especially Dobsonians, tend to lose collimation if they are not permanently mounted and are often moved. Fortunately, they are easy to collimate with a laser collimator. Other designs, such as the Schmidt–Cassegrain, are not so straightforward to collimate. So if you suspect you have a collimation problem, please see your observatory support staff.

The telescope you will be using is looking at the sky, which is considered a curved surface. Likewise, except for some systems that might have a flat secondary, our optical surfaces are also curved, yet our imaging device is flat. This mismatch of a curved surface projected onto a flat surface leads to **field curvature**. This leads to the distortion of separation distances between objects within the field as well as nonuniform illumination of the imaging surface. This is a particular problem for low F-number fast telescopes that have a large field of view. Field curvature can be addressed with additional optics, or in very rare cases, such as with the Kepler Space Telescope, by curving the detector surface. As we will see later, in the chapter on CCD, certain processing techniques can also partly resolve field curvature.

3.5 Introduction to Astronomical Detectors

For most of human civilisation, the best optical imaging system was the eye. The human eye has a lens and adjustable aperture and a light-sensitive film, the retina. In humans, the retina is a thin film on the back of the eye covered in light-sensitive cells. The rods are sensitive to all light, and the cones are colour sensitive. There are three types of cones, sensitive to red, green, and blue light only. Near the centre of the eye, at the focal point of the lens, we find the macula. Measuring only 5 mm across, it is where most of the cones are located. The inner part of the macula contains the fovea, where most detailed vision is found. If you are reading this book, you are doing so using your fovea. Outside the macula, the density of light-sensitive cells drops off, cones especially. Hence, humans have better light sensitivity in their peripheral vision. This leads to an observation technique whereby observers look at a faint target indirectly to get the light to impinge more fully on their rods than on their cones. Cones are much less sensitive than rods, hence the reasoning behind this technique. This is also the reason that you lose your colour vision in dark conditions; there isn't enough light for the cones.

As a detector, the human eye isn't perfect. It has a limited aperture size of about 7 mm, a fixed focal length, poor quantum efficiency, with only about two percent of photons generating a signal, and a fixed exposure time of about 30 s.

In 1840, the first image was taken of an astronomical object, the Moon, and the science/art of astrophotography was born. As improvements in sensitivity were made, scientists became able to image fainter objects than could be observed with the naked eye. The combination of long exposure times and greater sensitivity (photographic film is five times more efficient than the human eye) allowed a much deeper image to be taken than could be observed with just the eye. Film also allowed a permanent record of the observation that is not so strongly subject to human interpretation as hand-drawn images. It also enabled more accurate measurements of position and brightness. However, photographic film response to light is highly nonlinear, with most of the image being formed in the first few seconds of the observation, which places a limit on photometric observations. Additionally, the film has to be developed, which generates a considerable lag between the observation of the target and seeing the image produced. It was not uncommon to find a night's observation ruined by poorly handled film.

In the 1970s, professional astronomers started to use light-sensitive detectors using charged-coupled devices (CCD). We discuss CCDs (as well as the similar technology of the complementary metal-oxide semiconductor (CMOS) in later chapters. At their core, digital light detectors capture electrons displaced by being struck by photons in cells called pixels, and these pixels are turned into an image. Modern astronomical imagers make use of long exposure times, have exceptionally good quantum efficiency, and respond linearly to light throughout most of their range. Perhaps, however, their main advantage over photographic film is that they can be digitally measured and manipulated. It is easy to measure star brightness and position. Commercial digital cameras for terrestrial use employ both CCD and CMOS chips, although the move now is mostly towards CMOS, as they are cheaper to produce and have lower power consumption while providing performance approaching that of CCDs. A very good digital single lens reflex (DLSR) body can cost several thousand dollars, whilst a top-end amateur astronomical imaging camera can cost ten times as much while offering what appears to be less functionality. What you might find more surprising is that the imaging chips in both camera types might well have come out of the same factory and even the same production run. So why is the astronomical camera so much more expensive? The first reason is quality control. Every imaging chip is tested. With millions of pixels, likely a few will not work or underperform. For normal use, this is not a problem, but for use in astronomy, it is, and hence only the very best imaging chips go into astronomical cameras. An additional problem is the noise generated by the circuitry attached to the imaging chip. Although for non-astronomical use, low-noise electronics are desirable, they is very costly, and there is a cost–benefit tradeoff. In astronomical imagers, low-noise readout of imaging chips is vital, and custom-designed electronics with carefully selected and tested components are used to achieve this goal, driving up cost. Typical research-grade images have read noise an order of magnitude less than that of a typical non-research camera. Lastly, there are benefits of large-scale production to be considered. Companies such a Nikon, Sony, and Canon may produce hundreds of thousands of exemplars of any one model, whilst an astronomical camera manufacturer may produce only a few hundred.

When you start taking images, it will most likely be with an astronomical camera. You might also be using an adapted DLSR, however, and it is important to keep in mind that DLSRs, although they can produce amazing astronomical images, are not suitable for producing images from which scientific measurements can be made, as they have inbuilt nonstandard filters and perform much of the calibration internally without human intervention.

In recent years, astronomical webcams have become popular in the amateur community and have been moving slowly into professional use. Initially, these where domestic webcams hacked to be used for astronomy by the removal of their infrared filter, the attachment of a tube to allow fitting to a telescope, and the use of special software. Nowadays, astronomical webcams, both converted versions of standard webcams and webcams built specifically for astronomy, can be purchased off the shelf. Webcams have, in general, low sensitivity and high noise, as they are designed for video rather than imaging. However, through a technique known as lucky imaging, it has become possible to achieve images of very high resolution of bright astronomical objects, for example the Moon and the planets. The detailed use of lucky imaging is discussed in later chapters.

Chapter 4
Time

Abstract The measurement of time was one of the first applications of astronomy, and many of observational astronomy's underpinnings are related to time. Knowing the Julian date and local sidereal time is key in identifying what objects can be observed and when.

4.1 Introduction

Our modern lives are ruled by time; either the "tick tick tick" of the clock or, more commonly now, the fast harmonic hum of a quartz crystal. We work to timetables and deadlines, and the term "working day" often involves little actual daylight. However, for the greater part of human existence, when cultures were predominantly hunter–gatherer, pastoral nomad, and agrarian, time was merely the changing of the seasons and the rising and setting of the Sun. When to eat, when to sleep, when to move the herds, when to plant, and when to harvest. Our ancestors noticed that the point of sunrise on the horizon changed through the year. When the day was at its shortest, the Sun rose at its furthest point north; when at its longest, the Sun rose at its furthest point south; and twice a year, when the days were equal in length, the Sun rose at the midpoint between the northern and southern extremes. The first two events are known respectively as the winter and summer solstices. The term comes from the Latin for Sun (*sol*) and the verb to stand still (*sistere*). The latter pair are the spring and autumnal equinoxes, from the Latin *aequalis* (equal) and *nox* (night). Great significance was placed on these events, and great structures, such as Stonehenge in Great Britain, were erected to observe them. These structures became, in effect, the first calendars.

Later, with the development of cities, military forces, organised religion, and elite ruling classes, marking the passage of time and its measurement became increasingly important to intrasocietal coordination. Within an extended community, the populace needed to know when to pray, when to change the guard, and when the tax collector was coming. Time was measured using sundials, water clocks, sand timers, and

M. Gallaway, *An Introduction to Observational Astrophysics*,
Undergraduate Lecture Notes in Physics,
https://doi.org/10.1007/978-3-030-43551-6_4

marked candles. The year was often, but not always, broken into 365 days, with subdivisions, now known as months, based on the cycle of the Moon.

In 46 BCE, the introduction of the Julian calendar by Julius Caesar as the official Roman calendar resolved many of the problems of earlier systems of timekeeping, and the span and influence of the Roman Empire spread the Julian calendar throughout Europe and North Africa.

By 1582, it was becoming clear that small discrepancies in the Julian calendar were causing the seasons to drift. This was resolved by the introduction of the Gregorian calendar, named after the pope at the time, which resulted in a ten-day offset. The change encountered considerable resistance, with some people believing they were losing ten days of their lives! Surprisingly, it was not until the 1920s that the last country using the Julian calendar finally abandoned it.

It is worth stopping for a brief while before we move from date to time to discuss BCE and CE (also known as BC and AD). The move to a CE dating system occurred before the introduction of the concept of zero in the Western world. The consequence of this is that 1 BCE is followed by 1 CE. In general, this should not cause you any problems unless you are using dates that predate the common era. Likewise, you should be aware that there is a ten-day difference between Julian and Gregorian dates.[1] If you are using a pre-Gregorian date for some reason, you will need to convert.

By the fourteenth century, mechanical clocks, used mainly for religious reasons and important cultural gatherings, were becoming increasingly common. The introduction of the pendulum by the Dutch astronomer, mathematician, and inventor Christiaan Huygens and the hairspring by Robert Hooke, who coincidently also coined the word "cell" in relation to biology, led to clocks becoming more widespread, portable, and affordable.

All clocks have inaccuracies and require resetting to a known time on occasions. Up until 1847, the vast majority of clocks were set at noon. However, the timing of noon is a function of local longitude, and hence noon in London occurs approximately ten minutes before noon in Bristol, which lies $2° 3''$ west of London. It was not until the coming of the railways, requiring well-coordinated timetables over a large network, that the need for a centrally set time system become apparent. Ultimately, this led to the introduction of time zones and **Greenwich Mean Time** (GMT) as the de facto standard.

Typically, time zones are approximagely one hour wide and are measured from the time in Greenwich, in the United Kingdom, **Greenwich Mean Time** (GMT), with each hour representing $15°$ of latitude. Hence Western European Time, one hour ahead of GMT, is notated GMT+0100, whilst Honduras is at GMT-0600. Although since 1972 the standard has been **Coordinated Universal Time One** (UT1), the difference between GMT and UT1 is small enough (0.9 s) that for most purposes in astronomy it can be ignored. The introduction of daylight saving time (British summer time in the UK) can somewhat complicate matters. Beware if you are not

---

[1] The two calendars continue to diverge. The difference increased to 11 days in 1700, and today it is 13 days.

using GMT or a standard GMT time-zone-related offset. Best practice in astronomy is not to use daylight saving time. If you need to note down time in your notes, make sure that you use GMT or your local equivalent.

4.2 Solar Time

Johannes Kepler was a seventeenth-century astronomer and astrologer (at the time, the two were indistinguishable) born in Weil der Stadt, a town in what is now southern Germany. He worked for some time with Tycho Brahe, one of the greatest observational astronomers of the time. Tycho was a colourful character who lost his nose in a duel over a mathematical problem (both sides turned out to be incorrect), had a pet elk, which died when it fell down a set of stairs whilst drunk, was one of the richest people in Europe, and died under suspicious circumstances, possibly poisoned by his assistant over a woman. Claims that he died when his bladder burst when he refused to leave a banquet in order to relieve himself are likely to be untrue. Kepler used Tycho's excellent observations of the motion of the planets to formulate his famous three laws of planetary motion. It wasn't until after Kepler's death that Newton's three laws and his law of universal gravitation led to a theoretically grounded understanding of the phenomena described by Kepler's laws.

Kepler stated in his three laws that each planet orbits the Sun in an ellipse, with the Sun at one of the foci; that a line drawn between the Sun and the planet will sweep out equal areas in equal times; and that the square of the orbital period of the planet around the Sun is proportional to the cube of the semimajor axis of its orbit.

At this point, you may be wondering what this has to do with time. If we measured, over the course of a year, the time at which the Sun passes through the line drawn from the northern cardinal point, over the observer's head, to the south cardinal point, the **meridian**, we find that it is not the same every day. Some days, it will occur before noon on our clock, sometimes after; and on four days, it will be close to the same time. The variation is between plus fifteen and minus fifteen minutes. This difference between the time as measured by your watch, **mean solar time**, and the time indicated by the Sun, **apparent solar time**, and the translation between the two is known as the **equation of time** (see Fig. 4.1).

There are two reasons for the difference between mean solar time and apparent solar time. Firstly, let us consider Kepler's first and second laws. The first law tells us that for half the year, the distance between the Earth and the Sun decreases until Earth reaches its closest approach, **perigee**, occurring in January. It then increases until reaching maximum distance **apogee** in June. From the second law, we know that the Earth must be moving faster in its orbit, the closer it is to the Sun. Hence, the Earth will travel a greater angular distance around the Sun in a day at perigee than at apogee, resulting in the Sun moving slightly ahead of the mean. Likewise, because the Earth is slightly closer to the Sun during perigee than at apogee (and the amount is very small, only 1.6%), the apparent movement of the Sun caused by the Earth's orbit is larger.

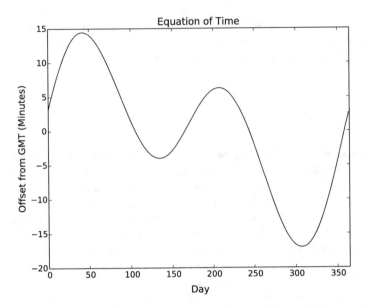

Fig. 4.1 Figure showing the difference, over the course of a year, between 12:00 mean solar time and the time the Sun crosses the local meridian

Turning to the second cause. You will no doubt be aware that the Earth's axis is inclined by 23.5° to its orbital plane and that it is this that accounts for the seasons. The Sun tracks along the ecliptic every day, with the peak of its diurnal movement increasing daily until the summer solstice, at which point it begins to descend, reaching its minimum at the winter solstice. Looking at Fig. 4.2, which projects the Sun's position on the ecliptic onto the equator, we see that the Sun tracks across the sky at a set angular rate, governed by the rotation of the Earth. Near the summer solstice, the height of the Sun at noon is at its highest. Hence the Sun has further to travel, and once its position is projected onto the horizon, it is behind that of a Sun that would not have an inclined path.

It is the combination of these two effects, the eccentricity of the Earth's orbit and its inclination to the ecliptic, that is responsible for the equation of time.

4.3 Julian Date

Astronomers do not use standard Gregorian calendar dates. Instead, a decimal year-independent system is used, giving what is known as the **Julian date** (JD). The Julian date is the number of days since noon on 1 January 4713 BCE by the Julian calendar, a date chosen because it is the convergence of the indiction, metonic, and solar cycles, an event that will not occur again until 3268 CE. Fractions of days are expressed as decimals; hence 6 h is 0.25 of a day. The calculation from Gregorian date to JD is

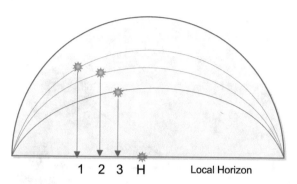

Fig. 4.2 Figure showing the effect of the inclination of the Earth's orbit on the equation of time. Note the three arcs representing three different days, with the position of the Sun on each arc after a set time and that position projected onto the horizon. As can been seen as we move towards the summer solstice, the apparent solar time, as shown by the projection onto the horizon (1, 2, and 3), falls behind mean solar time, which is indicated by H on the x-axis

straightforward but long and laborious, which makes it an ideal candidate for a small computing project. However, it can be calculated in a truncated form using (4.1) for dates after 1 January 2000. For (4.1), Y is the current year, D is the number of days since the start of the current year,[2] and L is the number of leap days since 2001. For fractions of a day, take the number of fractional days since midday on the day you are working on and add it to the Julian date found using (4.1). Hence 6 a.m. is −0.25, whilst 6 p.m. is +0.25.

The reader should be aware that a number of different epochs are used for the Julian date other than noon on 1 January 4713 BCE. For example, truncated JD uses midnight on 24 May 1968, and Dublin JD uses midday on 31 December 1899, so be aware what system you need to use and ensure that your use is consistent. If you are looking at a Julian date that predates the introduction of the Gregorian calendar, I recommend using a long-form equation rather than (4.1).

$$JD = 2451544.5 + 365 \times (Y - 2000) + D + L \qquad (4.1)$$

4.4 Sidereal Time

A mean solar day, the time in between consecutive solar transits from east to west, is 24 hr in duration. This accommodates the approximate one degree per day displacement of the Sun caused by the Earth's orbital motion (see Fig. 4.3). However, we use a 24-h clock in our day-to-days lives. Hence **sidereal time**, the time between consecutive stellar meridian crossings, diverges from solar time at a rate of approximately 4 min per day, with the exact difference determined by a number of effects in

---

[2]Note that most spreadsheets find this very quickly.

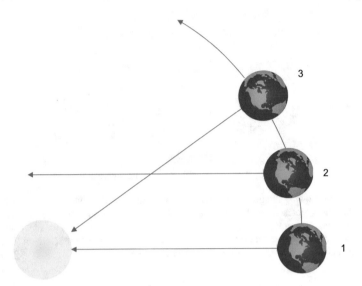

Fig. 4.3 Illustration showing the difference between solar and sidereal time. At position one, the Sun is directly overhead. At position two, the Earth has made one full rotation, but due to its orbit around the Sun, the Sun is not now overhead. At position 3, the Earth has had to rotate a further four minutes for the Sun to return to overhead

addition to the difference between the lengths of the solar and sidereal days. Given that the **local sidereal time** (LST) is the right ascension (RA) of an object crossing the meridian at that moment, we can see that an observer needs to know the local sidereal time for when they intend to make their observations. In order to convert from the RA and DEC (declination) used to locate the object on the celestial sphere to the AltAz used to point the instrument.

In order to find the local sidereal time, we need first to find the current GMT. You should be aware whether your local time zone has adjusted to daylight saving time. Let us use an observer in Boston (Massachusetts, not Lincolnshire) as an example. Boston is in UTC-5; hence when it is 14:00 in Boston, the GMT is 19:00, since $GMT = LT - TimeZone$.

By finding the number of days D since midday on 1 January 2001, where partial days are expressed as fractions, we can apply (4.2) and (4.3) to determine the Greenwich sidereal time. Reversing the operation we undertook to find GMT gives us our local sidereal time:

$$t = 18.697374558 + 24.06570982441908 \times D \qquad (4.2)$$

and

$$GMST = (t/24) - int(t/24). \qquad (4.3)$$

For example, we wish to know at what time the star Vega crossed the meridian from the viewpoint of an observer in Boston on 7 July 2015. Boston is at $43°21' 29''$ N $71°3' 49''$ W and Vega is located at $18^h 36^m 56^s +38° 47' 01''$. Hence, the transit occurred at approximately 18:37 LST. We need to find the GST for that LST, which we do as follows:

1. Convert the LST to decimal hours. So $18^h 36^m 56^s$ becomes 18.61555556.
2. Convert the longitude into decimal degrees. So $71°3' 49''$ W becomes $71.06361111°$W.
3. Covert the longitude to hours by dividing by 15, to give 4.737574074.
4. If the location is west of Greenwich, add the longitude to the LST; if east, subtract it. If the result exceeds 24, subtract 24; if it is negative, add 24. Hence our decimal GST is 23.35312963, or $23^h 21^m 11^s$.

Now that we know the GST of the transit, we need to find UT for $23^h 21^m 11^s$, GST on 7 of July 2015. For this, we need the Julian date at 0^h, which is found as follows:

1. Denote the year, month, and day by y, m, and d respectively. Hence, 7 July 2015 becomes $y = 2015$, $m = 7$, $d = 7$.
2. If $m \leq 2$, then subtract 1 from y and add 12 to m.
3. If the date is greater than 14 October 1582, find $A = INT(y/100)$ and $B = 2 + A + INT(A/4)$. Hence $A = 20$ and $B = -13$. If the date is earlier than 15 October 1582, then $B = 0$.
4. If $y < 0$, then $C = INT(365.25 \times y) - 0.75)$. Otherwise, $C = INT(365.25 \times y)$. Hence $C = 735978$.
5. $D = INT(30.6001 \times (m + 1))$. So $D = 244$.
6. The Julian date is then $JD = B + C + D + d + 1,720,994.5$, or for 7 July 2015, 2,457,210.5.

We can now find the UT for the GST we found above.

1. Calculate $S = JD - 2,451,545.0$. $S = 5665.5$.
2. Calculate $T = S/36,525.0$. $T = 0.155112936$.
3. Calculate $T_0 = 6.697374558 + (2400.051336 \times T) + (0.000025862 \times T^2)$. $T_0 = 378.9763853$.
4. Reduce T_0 to the range 0–24 using $T_0 = T_0 - INT(T_0/24) \times 24$, $T_0 = 18.97638528$.
5. Convert GST to decimal, which we already have: 23.35312963.
6. Subtract T_0 from the decimal GST, adding or subtracting 24 as appropriate, which gives 4.376744348.
7. Multiplying by 0.9972695663 gives the decimal UT, i.e., 4.364793938, or $4^h 21^m 53^s$ UT.

Boston's local time is UTC-5, so the local UT is $23^h 21^m 53^s$. As it turns out, this is the transit for the previous evening. However, we can safely assume that the transit will be around the same time. Otherwise, we will have to perform the calculation.

Chapter 5
Spheres and Coordinates

Abstract In observational astronomy, we are dealing with the celestial sphere, which is non-Euclidean. Hence, astronomers use the non-Cartesian Alt-Az and equatorial coordinate systems to point telescopes and identify objects. The non-Euclidean nature of our observations means that we have to use spherical geometry when measuring the position of objects in the sky.

5.1 Introduction

For thousands of years, people thought of the sky as a hollow crystal sphere with the Earth at the centre to which the stars were attached. Although we know that this concept is incorrect, the idea is not entirely without its merits. The stars do not change position relative to each other over a human lifetime, and they appear as point sources, with no parallax discernible to the human eye. The consequence of this is that stars appear in the same position in the sky relative to each other for all observers on Earth. We can, therefore, make a representative map of the stars as seen from Earth using a sphere with the stars placed on its surface and the observer located at its centre, with the Earth's equator and meridian lines projected onto the sphere. This sphere is known as the **celestial sphere**.

For an observer, the current position of a stellar object in the visible sky is governed by several factors: the location of the observer on the Earth, the time, and the position of the object on the sphere of fixed stars in relation to established reference points. All of these depend on spherical geometry.

Let us consider a sphere. Extend a line of diameter vertically, both up and down, through the surface of the sphere. For rotating bodies such as the Earth, we make this the axis of rotation. For purposes of orientation we shall call the uppermost point of interception the **north pole** and the lower the **south pole**, although this is entirely arbitrary. We draw a circle intercepting both the north and south poles so that this circle has the same circumference as the sphere. Such a circle is known as a **great circle**, and if we used it to bisect the sphere, it would do so through the centre. We shall call this great circle the **meridian**. We now draw a second great circle perpendicular

© Springer Nature Switzerland AG 2020
M. Gallaway, *An Introduction to Observational Astrophysics*,
Undergraduate Lecture Notes in Physics,
https://doi.org/10.1007/978-3-030-43551-6_5

Fig. 5.1 Figure showing the position of the fundamental plane and the meridian in relation to the poles

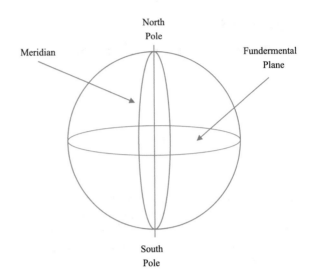

to the meridian. This second great circle is the **equator**, although when used within a coordinate system it is more correctly known as the **fundamental plane**; see Fig. 5.1. Any circle on the surface of the sphere that does not bisect the centre of the sphere is known as a **small circle**.

One of the two points where the fundamental plane intercepts the meridian becomes the origin for the coordinate system using that fundamental plane. An object can be located on the surface of this sphere using just its angular distance from the meridian line and its angular distance above or below the fundamental plane.

The most commonly encountered such arrangement is the familiar longitude and latitude system for determining position on the Earth, with the two great circles being the equator and the prime meridian running through Greenwich Observatory, in London. Hence, the origin is the point at which the prime meridian intersects the equator, located in the Atlantic Ocean off the coast of Equatorial Guinea. Typically, longitude and latitude are listed in degrees, minutes, and seconds north or south of the equator and east or west of the meridian. However, in some cases, you may find that decimal degrees are used or that the cardinal points are replaced with a plus (for north or west) or a minus (for south and east). In some cases, for example in working out sunrise or sunset times, you will need to know the observatory's altitude. Although altitude is typically taken from mean sea level, global positioning systems (GPS) use true spherical coordinates in order to determine position. Therefore, the reported altitude is based on the distance from the centre of the Earth, with the Earth's radius taken as 6,378.137 km. Additionally, the GPS system does not translate perfectly into geographical systems, although this is unlikely to cause any problems for an astronomical observer, as the error can be less than 100 m. You should be aware of this if the telescope you are using uses a GPS receiver to determine time and position. If you have a handheld GPS, you may notice in the setting that it uses a coordinate system known as WGS84, for World Geodetic System 1984.

5.2 Altitude and Azimuth

The simplest coordinate system you will encounter in astronomy is the altitude and azimuth system, also known as the horizontal system, but more commonly known as AltAz. AltAz (or Alt-Az) is based on the two great circles formed by the observer's horizon and celestial meridian line. The celestial meridian line passes through the **zenith**, the point directly above the observer, and the **nadir**, the point directly below the observer, intersecting the north and south cardinal points. For an observer in the northern hemisphere, the origin is the point on the horizon due south of the observer. For observers in the southern hemisphere, it is the point on the horizon due north of the observer.

The AltAz system is determined by factors local to the observer, and it is not fixed to the celestial sphere. Hence, astronomical objects transit through the AltAz grid during the course of the night as the Earth rotates. However, it is the AltAz grid that is used to point a telescope, and the observer must translate between the coordinate system of the target and AltAz. For an observer in the northern hemisphere, an object is located using its azimuth (A), the object's angular position along the **observer's horizon**, measured eastwards from the most northerly point on the horizon, and its altitude (α), the angular elevation above the horizon. Hence, for an observer in the northern hemisphere, an object lying due south on the horizon has an Alt$=0°$ and Az$=180$ °, whilst the point lying due north on the horizon has Alt$=0°$ and Az$=0$ °. A star directly overhead, i.e., at the zenith, would have Alt $= +90°$, while its Az could be anything at all, as azimuth angles converge at this point. For observers in the southern hemisphere, it would be an object to the north on the horizon that would have Alt $= 0°$ and Az $= 0°$, whilst the object at the zenith would have an Az of $-90°$.

As AltAz is a local system, it is used to determine not only when and where to point a telescope but also whether the target is visible, whether it is in its optimal position in the sky for observing, and whether the location of the target is going to exceed the telescope's limits of orientation. In many cases, the instrument you will be using will have some limitation as to where it can point. This is typically the minimum altitude that can be observed, which is rarely as low as 0°, as often there is some obstruction, for example the edge of the dome or trees, that prevents this. Furthermore, with the increasing trend of ever larger astronomical cameras, maximum altitude limits have also become more prevalent due to the inability of these larger cameras to pass safely under the telescope forks. Although somewhat rarer than altitude limits, some sites also have azimuth limits, due, for example, to physical obstructions, problems with prevailing winds, or mount limitations. In many cases, you may find that a telescope has declination limits. Typically, such limits occur when the telescope is **polar aligned** (when the Az axis of the telescope points at the pole), as most are. Declination limits are similar to altitude limits but are adjusted for the tilt in the telescope's vertical axis due to the polar alignment. Before you observe, you should be aware of the telescope's pointing limits whether

Fig. 5.2 Figure showing the alignment of the AltAz system. Altitude is measured in degrees from the horizon toward the zenith, while Azimuth is measured from due north

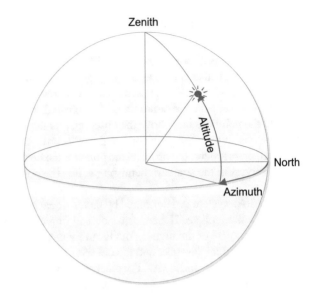

in altitude, declination, or azimuth and the restraints it may put on your observations (Fig. 5.2).

The axis of the Earth's rotation is tilted by approximately 23.5° in reference to the plane of its orbit around the Sun. Consequently, the Sun, the planets, and the stars all appear to follow arcs across the sky as the Earth rotates. Along with the Earth's orbit, those of the other planets all lie in approximately the same plane. Having their motions thus confined, when viewed from the Earth they appear to follow a common path across the sky. The line to which they appear to be affixed, the **ecliptic**, is inclined by 23.5° to the Earth's equatorial plane, as explained by its axial tilt. This line passes through the 12 **zodiacal** constellations, and for reasons to be explained later, Ophiuchus now also straddles the ecliptic. The ecliptic intercepts the celestial equator at two points: the vernal equinox and autumnal equinox. The days at which the Sun is seen rising at these points mark the start of astronomical spring and autumn respectively, in addition to being the point of time where day and night are of equal length, whence the name. The solstices are the two points where the Sun rises farthest north and south (and hence the longest and shortest days); they mark the start of astronomical summer and winter, respectively.

5.3   Equatorial Coordinate System

The equatorial coordinate system is the most common coordinate system in astronomy. It uses the Earth's equator projected onto the celestial sphere—forming the celestial equator—as its fundamental plane, and the great circle passing through the vernal equinox (also known as the first point of Aries, which for reasons that will become clear is no longer in Aries) and the north and south celestial poles to determine the origin.

The equatorial system identifies an object by its right ascension (RA or α), which is the angular distance eastwards along the celestial equator between the object and the first point of Aries, and its declination (Dec or δ), which is the angular separation between the object and the celestial equator. Objects lying on the celestial equator have a declination of $0°$, with those north and south of the line having positive and negative values respectively. Typically, right ascension is measured in hours, minutes, and seconds, and declination in degrees, minutes, and seconds, that is, in **sexagesimal notation**. You should be aware that minutes and seconds in RA are not of the same length as minutes and seconds in Dec. If you are performing positional measurements, you may wish to to resolve this problem by converting your RA and Dec to decimal notation.

You should be aware that on occasion, decimal degrees may be required or supplied. In general, decimal degrees are given as just a decimal number, while sexagesimal notation has the individual components separated by a space, by hms (for hours, minutes, and seconds), or more commonly, by a colon. Hence the star known as Betelgeuse, or α Orionis, has an RA and Dec denoted by 05 h 55 m 10 s +07 d 24 m 25 s or 05:55:10+07:24:25 or, in decimal notation, 238.7917+7.406944.

The Earth's axis is precessing, like a slowing spinning top. Consequently, over long time periods, points that seem fixed gradually reveal themselves to be in motion. For example, the celestial poles appear to describe circles. It is for this reason[1] that the first point of Aries is no longer is Aries and also why there are now 13 zodiacal constellations. Over short periods, nutation does not pose a problem for astronomers. However, over longer periods stars will move away from their catalogue positions due to the rocking of the Earth's axis, a process known as **nutation**. In an attempt to resolve this problem, astronomical catalogues have fixed dates for the positions they represent, known as the **epoch**. The current epoch is J2000; it uses positions determined from midday 1 January 2000. The previous epoch was B1950, which was in use prior to 1984. Although you should encounter only J2000 coordinates, if you are using an older catalogue or academic paper, you may come across B1950 or even older epochal systems. In such cases, you should endeavour to find the object's J2000 coordinates, using, for example, the SIMBAD astronomical catalogue. You may also encounter JNow when you are using astronomical software, which gives an object's RA and Dec based on its J2000 positions, the current date, and the precision rate. In general, users should avoid using JNow, as at the time of writing the differences between J2000 and JNow are minimal. However, as J2000 becomes increasingly outdated, JNow may become necessary (Fig. 5.3).

For an observer in the northern hemisphere, stars with declinations greater than $\theta - 90°$ (where θ is the observer's latitude) should be visible at some point during the year, although that depends on the quality of the local horizon. Likewise, in the southern hemisphere, stars with declinations less than $\theta + 90°$ will be visible.

---

[1] As well as others that are beyond the scope of this book.

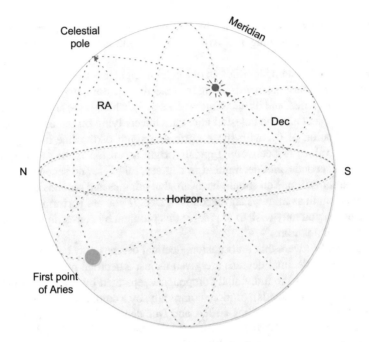

Fig. 5.3 Figure showing the alignment of the equatorial coordinate system

This leads to an important observation. If not all stars are visible to an observer, it follows that some are always visible. Such stars are known as **circumpolar** stars.[2] Stars are circumpolar in the northern hemisphere if $\theta + \delta \geq 90°$, and in the southern hemisphere if $\theta + \delta \leq -90°$. So for an observer in Greenwich, where $\theta = 51.48°$, the star Vega $\delta = 38.8°$ is just barely circumpolar.

Another set of great circles comprises those that run perpendicular to the equator and therefore pass through the poles. Hence, every point along the line between the two poles and on the same side of the sphere have the same right ascension. This line is known as an **hour circle**. The meridian is a special member of the set of hour circles, being the only one that passes through the origin. The angular distance between the hour circle on which an object currently lies and the meridian is the **hour angle**, which is simply the local sidereal time minus the RA of the object.

5.4 Galactic Coordinates

In 1958, in part to overcome the problem with precession, the **International Astronomical Union (IAU)**, the de facto governing body for astronomy, introduced the galactic coordinate system. In this system, the origin is the galactic centre, which

---

[2]Technically, stars that are never visible are also circumpolar, but this term is hardly ever used in this context.

is marked by the radio source (and supermassive black hole) Sagittarius A*. The north galactic pole is designated as having a latitude (b) of $+90°$, the south galactic pole is designated as having a latitude of $-90°$ with longitude (l) measured in degrees with a right-handed convention along the galactic plane, with galactic center at $l = 0°$ and anticentre at $l = 180°$. Hence in this system, Betelgeuse is located at 199.7872−08.9586.

Galactic coordinates are most commonly used by galactic astronomers, those whose interests lie within the galactic plane. For extragalactic uses, the system is without merit, and in general, the equatorial system is used in these cases, although there also exists a supergalactic coordinate system for extragalactic use.

Because the system uses the galactic centre as its point of origin, the rate of precession is very small, since the orbital period of an object around the galactic centre is on the order of 200 million years. For high **proper motion** objects, the proper motion being the movement of stars against the more distant background, the rate of change is as great as it is for equatorial coordinates. However, these objects are normally extremely close, and so are unlikely to lie on the galactic plane, where galactic coordinates are used.

5.5   Other Minor Systems

An earlier system, now mostly used by those interested in objects within the **solar system**, is the **ecliptic coordinate** system. The fundamental plane for ecliptic coordinates is that defined by the ecliptic, the arc traversed by the Sun and, to some extent, the planets. Objects are located by their ecliptic latitude β and ecliptic longitude δ, with the origin at the first point of Aries.

The **supergalactic system** is a coordinate system for extragalactic objects. It uses a flat, two-dimensional structure formed by the distribution of nearby galaxies known as the **supergalactic plane** as its fundamental plane and its intersection point with the galactic plane as the origin. The system can be extended three-dimensionally with galaxies on the plane having an SGZ value of zero. In deference to galactic coordinates, supergalactic latitude is denoted by SGb, and supergalactic longitude by SGb.

5.6   Spherical Geometry

As you have seen, astronomical coordinate systems are based on spheres. However, the surface of a sphere is nonplanar, and therefore, the measurement and calculation of distances and angles between astronomical coordinates requires the application of spherical rather than planar geometry.

Perhaps one of the more day-to-day applications of spherical geometry is in the routes flown by aircraft. You may have noticed on intercontinental flights that the

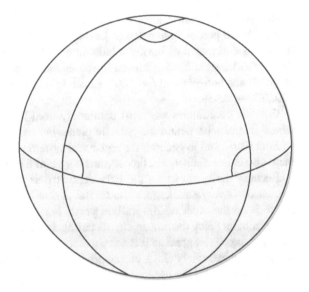

Fig. 5.4 Image showing the formation of a triangle on the surface of a sphere the sum of whose internal angles exceeds 180°

aircraft has a GPS location offering in the entertainment system, in which you can see that the course is not a straight line on the screen between the point of departure and the destination but is, in fact, an arc. In spherical geometry, a line is the shortest distance between any two points and, except in the case where those two points can be connected by a line passing through the centre of the sphere, that line will always be a segment of a great circle.

We are taught early in our mathematical education that the sum of the interior angles of a planar triangle is always 180°. Now let us consider a sphere and three great circles, one the equator, one the meridian, and one perpendicular to both the meridian line and the equator that I shall call the offset. These three circles form a triangle whose vertices are located at the pole where the meridian and the offset meet, the intercept of the equator and the meridian, and the intercept of the equator and the offset. As can be seen from Fig. 5.4, all these points have interior angles of 90°, a result that is, of course, impossible for a planar triangle. In fact, the interior angles of a triangle on the surface of a sphere are always greater than 180° and less than 540°; see Fig. 5.4.

In many cases, you will either be translating between coordinate systems or measuring the distances between two points. In general, given the amount of computation to convert between systems and the likelihood of error, I strongly recommend either the Coco programme contained within the StarLink package or the Python astropy.Coordinates package.

Measuring the distance between two points will become increasingly important as we progress through this book. In general, (5.1) is sufficiently precise for most of our needs. In this case, $\phi_1, \phi_2, \lambda_1, \lambda_2, \Delta\lambda$, and $\Delta\phi$ are the latitudes of points 1 and 2, the longitudes of points 1 and 2, the difference between the longitudes of 1 and 2, and the difference between the latitudes of 1 and 2, respectively, with $\Delta\sigma$ the central

Fig. 5.5 The above diagram shows three points A, B, and C on the surface of a sphere that are connected by the edges a, b, and c to form a *cap*

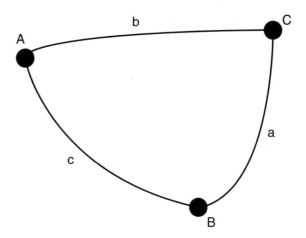

angle between points 1 and 2:

$$\Delta\sigma = arctan\left(\frac{\sqrt{(\cos\phi_2 \sin\Delta\lambda)^2 + (\cos\phi_1 \sin\phi_2 - \sin\phi_1 \cos\phi_2 \cos\Delta\lambda)^2}}{\sin\phi_1 \sin\phi_2 + \cos\phi_1 \cos\phi_2 \cos\Delta\lambda}\right).$$

(5.1)

In situations in which you are measuring the differential offset, i.e., when the offset from the origins is unknown or perhaps not to be trusted, a different approach may be taken. If, for example, we have two reference positions A and B and we wish to find the location of a point C in relation to points A and B, we can form a triangle on the sphere with edges AB, AC, and BC and internal angles A, B, and C. Such a triangle is shown in Fig. 5.5. In this situation, we can use (5.2), (5.3), and (5.4) and their derivatives to find the location of the unknown point.

$$\cos a = \cos b \cos c + \sin b \sin c \cos A \qquad (5.2)$$

$$\cos b = \cos c \cos a + \sin c \sin a \cos B \qquad (5.3)$$

$$\cos c = \cos a \cos b + \sin a \sin b \cos C \qquad (5.4)$$

As the distance between A, B, and C tends to zero, (5.2), (5.3), and (5.4) tend towards the sine rule (5.5). You may therefore in certain situations resort to (5.5), but be aware that it introduces further errors:

$$\frac{\sin A}{\sin a} = \frac{\sin B}{\sin B} = \frac{\sin C}{\sin c} \qquad (5.5)$$

Chapter 6
Catalogues and Constellations

Abstract Constellations in the night sky have been important to many cultures. Astronomers have adopted the Greek constellations to provide a regional map for the location of objects. Since before the invention of the telescope, astronomers have been classifying and cataloguing objects, so that there are now in excess of 200 such catalogues. Thus, knowing which catalogue is most appropriate to the task at hand is an important tool in observational astronomy.

6.1 Introduction

Astronomers like to order things, to find them, to identify what they are, and to put them into neat pigeonholes, perhaps not always successfully, as the ongoing problem with Pluto testifies. Clearly, with 200 billion stars in the Milky Way and 200 billion galaxies, not to mention the billions of solar system objects, nebulae, supernova remnants, X-ray sources, exoplanets, masers, radio sources, etc., there is a lot to catalogue.

Consequently, there is a very large number of astronomical catalogues (in excess of 200), some of which have considerable overlap and some that have a small number of members. This chapter is here to help you to identify an object type from a catalogue and highlight the more important and

6.2 Constellations, Asterisms, and Bright Stars

Many cultures have used the heavens to tell stories and illustrate their cultural mythologies. Stars were grouped to form images in a time when skies were darker, the night more intimidating, and imaginations perhaps more vivid than in more modern times. These imaginary figures became the constellations. The stars within the constellations are just random associations with no actual physical connection between members of any one constellation; however, they have an important role

© Springer Nature Switzerland AG 2020
M. Gallaway, *An Introduction to Observational Astrophysics*,
Undergraduate Lecture Notes in Physics,
https://doi.org/10.1007/978-3-030-43551-6_6

in both astronomy and the cultural identity of the civilisations that created them. Modern astronomers use the constellations as derived from Greek mythology with some additional constellations added at later dates that consist of the stars too far south to have been visible from the world of the ancient Greeks. It is also common, especially when speaking to non-astronomers, to use an anglicised version of the Greek name; hence the Great Bear is Ursa Major and the Swan is Cygnus.

The use of constellations may seem archaic, but they simplify the naming scheme of many astronomical objects in that they act in many respects as the county or state does as part of a postal address. Although there is no need to be able to identify all 88 constellations, as defined by the International Astronomical Union (IAU), by sight, you should be aware of the brighter and more prominent ones visible from your site, for example Ursa Major for northern observers, Crux for southern observers, and Orion and Leo for all observers. A full list of constellations can be found in Table 6.1, which also shows their official abbreviations and the hemisphere in which they are located.

Asterisms, like constellations, are collections of stars that often have no association beyond their location in a particular region of the celestial sphere. Technically, constellations are a subset of asterisms, those that they have been formally accepted as such, whereas most asterisms are simply a few of the brightest members of a constellation, a combination of constellations, or even a small segment of a constellation that possesses an easily identified structure. Hence in general, asterisms have far fewer constituents than constellations. Perhaps the most well known asterisms are the Plough (also known as the Big Dipper or the Saucepan), comprising the brightest seven members of the constellation Ursa Major, and the Summer Triangle, which is a northern hemisphere feature formed from the stars Deneb, Vega, and Altair, which are respectively the brightest members of Cygnus, Lyra, and Aquila.

If you are unlucky enough to be observing from a light-polluted site, it is probable that the asterisms formed from bright stars will be more prominent than the constellations. This will also be the case if you are setting up during twilight, when faint stars are still invisible. Bright stars are very important in observational astronomy. They are used to set up the alignment of GoTo telescopes,[1] as a check of alignment, a focusing target, and the first point of call when star hopping if you are not using a GoTo. It is unlikely that you will be able to identify the first visible star of the evening's observing (which might even be a planet), especially if you have been unable to observe for some time. However, as the sky darkens and more stars become visible, the asterisms will appear first, allowing you to identify a few stars and get set up before it is truly dark.

---

[1]GoTo telescopes have powered mounts and either a custom-built interface or computer control that allows the user to move rapidly between targets. Typically, the user has to point the telescope to three guide stars to set it up.

Table 6.1 The 88 constellations as defined by the IAU. Col. 1: Name; Col. 2: Abbreviation; Col. 3: Hemisphere located. Note that constellations listed as northern may be visible from the southern hemisphere and vice versa. Furthermore, some constellations extend across the equator

Name	Abbreviation	Hemisphere	Name	Abbreviation	Hemisphere
Andromeda	And	N	Lacerta	Lac	N
Antlia	Ant	S	Leo	Leo	N
Apus	Aps	S	Leo Minor	LMi	N
Aquarius	Aqr	S	Lepus	Lep	S
Aquila	Aql	N	Libra	Lib	S
Ara	Ara	S	Lupus	Lup	S
Aries	Ari	N	Lynx	Lyn	N
Auriga	Aur	N	Lyra	Lyr	N
Boötes	Boo	N	Mensa	Men	S
Caelum	Cae	S	Microscopium	Mic	S
Camelopardalis	Cam	N	Monoceros	Mon	N
Cancer	Cnc	N	Musca	Mus	S
Canes Venatici	CVn	N	Norma	Nor	S
Canis Major	CMa	S	Octans	Oct	S
Canis Minor	CMi	N	Ophiuchus	Oph	S
Capricornus	Cap	S	Orion	Ori	N
Carina	Car	S	Pavo	Pav	S
Cassiopeia	Cas	N	Pegasus	Peg	N
Centaurus	Cen	S	Perseus	Per	N
Cepheus	Cep	N	Phoenix	Phe	S
Cetus	Cet	S	Pictor	Pic	S
Chamaeleon	Cha	S	Pisces	Psc	N
Circinus	Cir	S	Piscis Austrinus	PsA	S
Columba	Col	S	Puppis	Pup	S
Coma Berenices	Com	N	Pyxis	Pyx	S
Corona Australis	CrA	S	Reticulum	Ret	S
Corona Borealis	CrB	N	Sagitta	Sge	N
Corvus	Crv	S	Sagittarius	Sgr	S
Crater	Crt	S	Scorpius	Sco	S
Crux	Cru	S	Sculptor	Scl	S
Cygnus	Cyg	N	Scutum	Sct	S
Delphinus	Del	N	Serpens [5]	Ser	N
Dorado	Dor	S	Sextans	Sex	S
Draco	Dra	N	Taurus	Tau	N
Equuleus	Equ	N	Telescopium	Tel	S
Eridanus	Eri	S	Triangulum	Tri	N
Fornax	For	S	Triangulum Australe	TrA	S
Gemini	Gem	N	Tucana	Tuc	S
Grus	Gru	S	Ursa Major	UMa	N
Hercules	Her	N	Ursa Minor	UMi	N
Horologium	Hor	S	Vela	Vel	S
Hydra	Hya	S	Virgo	Vir	S
Hydrus	Hyi	S	Volans	Vol	S
Indus	Ind	S	Vulpecula	Vul	N

6.3   Catalogues

Broadly, catalogues can be divided into stellar and nonstellar, although a small number are mixed. Additionally, catalogues can be further split into the historical divisions of visual, photographic, and digital observing, although a number of photographic surveys have now been digitised, which is a boon for those who wish to undertake proper motion studies requiring long gaps between observations.

You should be aware that many catalogues are geographically limited, in that they are taken with a single ground-based instrument that cannot see the entire sky. Likewise, not all objects capable of being detected by the survey will be in the catalogue, as some will always be missed. The number of detected objects compared to the number of expected detected objects from current computer models is the **completeness**. Most modern surveys have completeness levels in excess of 90%, but you should not be surprised if you find an object that is not in the catalogue you are using.

As stated above, the bright stars are extremely useful for anyone observing the night sky, and the ability to identify bright stars will greatly enhance the amount of observing time you will get at most teaching observatories, especially when things go wrong, as they often do. Unlike the names of the constellations, the common names of bright stars come mostly from the Arabic, with some Greek and anglicised names in use. Most telescopes equipped with GoTo controllers use the common name for bright stars.

Bayer Catalogue
Most astronomers use Johann Bayer's 1603 catalogue of approximately 2,500 of the brightest stars. Bayer used Greek letters and the constellation name, now often abbreviated, to identify each star in his catalogue, with alpha being the brightest, beta the second brightest, etc. Hence, α UMa is the brightest star in the constellation Ursa Major. Of course, there are far more stars in any constellation than there are Greek letters (in fact, there are approximately 10,000 stars visible with the naked eye), so this process applies only to the brightest, but it is very useful nonetheless. While we are on the subject of Bayer's catalogue, you should be aware that there are a number of errors in the ordering by brightness. For example, α and β Gem are the wrong way around. This error and others have never been corrected, and their existence is a topic of discussion within the astronomy community, as it is not clear whether such anomalies are truly errors or whether there has been an actual physical change in one or more of the stars. Whatever the reason, you should attempt to learn the common and Bayer names of the brightest stars visible from the location from which you intend to observe. A list some of these stars with their common names is presented in Table 6.2 along with their visual magnitudes.

Flamsteed Catalogue
Bayer's catalogue was improved upon with the introduction of a new catalogue by the first Astronomer Royal, John Flamsteed, in 1712. As with the Bayer catalogue, the Flamsteed catalogue is geographically limited and did not originally cover the

Table 6.2 Some of the brightest stars with their apparent magnitudes, Bayer identifier, and proper names

Magnitude	Bayer ID	Proper name	Magnitude	Bayer ID	Proper name
0.5	α Eri	Achernar	2	α Ari	Hamal
2.5	β Sco	Acrab	2.39	ε Boo	Izar
0.77	α Cru	Acrux	1.8	ε Sgr	Kaus Australis
1.51	ε CMa	Adara	2.08	β UMi	Kochab
0.6	β Cen	Agena,	1.96	δ Vel	Koo She
0.85	α Tau	Aldebaran	2.32	η Cen	Marfikent
2.44	α Cep	Alderamin	2.49	α Peg	Markab
2.01	γ Leo	Algieba	2.46	κ Vel	Markeb
2.12	β Per	Algol	2.28	α Lup	Men
1.9	γ Gem	Alhena	2.06	θ Cen	Menkent
1.76	ε UMa	Alioth	2.35	β UMa	Merak
2.15	γ And	Almach	1.68	β Car	Miaplacidus
1.74	α Gru	Alnair	1.3	β Cru	Mimosa
1.7	ε Ori	Alnilam	2.23	δ Ori	Mintaka
1.7	ζ Ori A	Alnitak	2.06	β And	Mirach
1.98	α Hya	Alphard	1.82	α Per	Mirfak
2.21	α CrB	Alphecca	1.98	β CMa	Mirzam
2.06	α And	Alpheratz	2.23	ζ UMa	Mizar
0.77	α Aql	Altair	2.17	γ Cen	Muhlifain
2.4	η CMa	Aludra	2.21	ζ Pup	Naos
2.37	α Phe	Ankaa	2.06	σ Sgr	Nunki
1.09	α Sco	Antares	1.91	α Pav	Peacock
−0.04	α Boo	Arcturus	2.43	γ UMa	Phecda
2.25	ι Car	Aspidiske	1.97	α UMi	Polaris
1.92	α TrA	Atria	1.15	β Gem	Pollux
1.86	ε Car	Avior	0.34	α CMi	Procyon
1.64	γ Ori	Bellatrix	2.1	α Oph	Rasalhague
1.85	η UMa	Benetnasch	1.35	α Leo	Regulus
0.42	α Ori	Betelgeuse	0.12	β Ori	Rigel
−0.72	α Car	Canopus	−0.27	α Cen	Rigil Kent
0.08	α Aur	Capella	2.43	η Oph	Sabik
2.27	β Cas	Caph	2.24	γ Cyg	Sadr
1.58	α Gem	Castor	2.05	κ Ori	Saiph
1.25	α Cyg	Deneb	1.86	θ Sco	Sargas
2.04	β Cet	Deneb	2.42	β Peg	Scheat
2.14	β Leo	Denebola	2.25	α Cas	Schedar
2.29	δ Sco	Dschubba	1.62	λ Sco	Shaula
1.79	α UMa	Dubhe	−1.46	α CMa	Sirius
1.68	β Tau	El Nath	1.04	α Vir	Spica
2.23	γ Dra	Eltanin	2.23	λ Vel	Suhail
2.4	ε Peg	Enif	1.78	γ Vel	Suhail
1.16	α PsA	Fomalhaut	2.39	γ Cas	Tsih
1.63	γ Cru	Gacrux	0.03	α Lyr	Vega
2.5	ε Cyg	Gienah	2.29	ε Sco	Wei
2.38	κ Sco	Girtab	1.84	δ CMa	Wezen

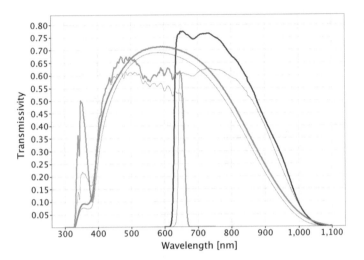

Fig. 6.1 Filter profiles of the Gaia filter set. Image courtesy of ESO

entire sky, although the catalogue has been extended after Flamsteed's death to cover the region too far south to be seen from Greenwich. Flamsteed used a similar naming method to that used by Bayer, but he used numbers instead of Greek letters, which significantly increased the number of names available. He also labeled stars in order of right ascension rather than magnitude. Hence the star known as Betelgeuse is known as α Ori in the Bayer system and 68 Ori in Flamsteed's.

Bonner Durchmusterung Catalogue
The last great visual catalogue was that undertaken by the Bonn Observatory between 1799 and 1875 and extended in 1886 by Cordoba Observatory to cover the regions too far south to be observed from Bonn. The Bonner Durchmusterung survey, as it is now known, contains approximately 325,000 entries, down to magnitude 9.5. The catalogue is divided into a series of strips of width 1°, with stars numbered within that strip in order of their right ascension. Hence Betelgeuse is known as BD +07 1055 in the Bonner Durchmusterung catalogue.

Gaia Archive
Gaia is an observatory of the European Space Agency designed to map the Milky Way to high astrometrical and photometric precision. Launched in 2013 and not due to finish its mission until 2022, at the time of writing there were 1.6 billion objects in its publicly accessible archive. Gaia has photometric uncertainties in the millimagnitude range and positional uncertainties in the range of hundredths of milliarcseconds. Gaia utilises a filter set that was at the time unique. As a result, users of the Gaia archive should exercise caution when comparing magnitudes to other surveys. The Gaia archive can be found at https://gea.esac.esa.int/archive/ (Fig. 6.1).

Messier Catalogue

Charles Messier was an eighteenth-century astronomer with a passion for comets. Unfortunately, eighteenth-century optics were not of the quality we now enjoy, and a very large number of non-comet objects looked like comets to Messier, so he created a catalogue of 110 objects in order to avoid later confusion. This catalogue is the now famous Messier catalogue. Given that Messier was operating from France, many of the objects are not visible from the southern hemisphere. Also, given that some of the objects, such as M31 (the Andromeda galaxy) and M45 (the Pleiades) are very well known naked-eye astronomical objects, it is unlikely that all of the objects in the list would be mistaken for comets. However, the 110 objects in the Messier catalogue are favourites among amateur astronomers for being bright, spread out over the sky, and comprising a mixture of **deep sky** astronomical object types, objects too faint to see with the naked eye and external to the solar system.

Caldwell Catalogue

Compiled in the 1990s by famed amateur astronomer and broadcaster Sir Patrick Moore (or more correctly Patrick Caldwell-Moore), the Caldwell catalogue is a list of deep sky objects omitted from the Messier catalogue due either to error, deliberate omission, or the object being too far south. The Caldwell catalogue contains 109 objects, including star clusters, planetary nebulae, and galaxies.

United States Naval Observatory Catalogues

The United States Naval Observatory has produced a series of catalogues, some of which are now considered obsolete. USNO A2.0 is the re-digitisation of the plates from three previous catalogues. USNO-A2.0 contains in excess of 500 million objects with R and B magnitudes down to magnitude 20. USNO A2.0 has now been largely replaced by USNO-B1.0, which contains in excess of one billion objects down to magnitude 21. Both USNO-A2.0 and USNO-B1.0 are accessible via SAO DS9.

UCAC2 and its replacement UCAC3 are high-precision (with regard to astrometry) catalogues to magnitude 16 in band f created from CCD images taken at Cerro Tololo Interamerican Observatory (CTIO) in Chile and the Naval Observatory Flagstaff Station (NOFS) in Arizona. The catalogue includes B, R, and I magnitudes from the SuperCOSMOS survey. Both UCAC3 and UCAC2 are accessible via SAO DS9.

The Naval Observatory Merged Astrometric Dataset (NOMAD) is a merged catalogue using the sources found within the Hipparcos, Tycho-2, UCAC2 USNO-B1.0, and 2MASS surveys. Hence NOMAD covers the B, V, R, J, K, and H bands down to approximately magnitude 20, depending on the source catalogue.

Hipparcos and Tycho

Hipparcos is a European Space Agency Telescope launched and operated between 1989 and 1993 with the goal of producing high-precision proper motions and parallaxes to stars within 200 parsecs. Observations were made in the B and V bands with a limiting magnitude of 12.4 in V, resulting in a 118,218-source catalogue. Objects in Hipparcos are numbered and prefixed with HIP; hence α Ori is HIP 27989 in the Hipparcos catalogue. The Hipparcos catalogue is being superseded by the Gaia catalogue with an order of magnitude more objects in more bands and to fainter magnitudes.

Two further catalogues were derived from the Hipparcos observations. Tycho-1 contains approximately 1 million entries down to a V magnitude of 11.5, while Tycho-2, which supersedes Tycho-1, contains 2.5 million objects down to a limiting magnitude similar to that of Tycho-1. Tycho object numbers are prefixed with TYC and consist of three numbers separated by a dash. The first number is the Hubble Guide Star Region; the second is the star number within that region, and the last is the component identifier for multiple star systems. Hence α Ori is TYC 129-1873-1 in Tycho-2.

The Hubble Guide Star Catalogue

The Hubble Guide Star Catalogue (GSC) was designed as the pointing catalogue for the Hubble Space Telescope and has been upgraded a number of times, with the current version at the time of writing being GSC-2, which is intended to be the pointing catalogue for Hubble's successor, the James Webb Space Telescope. The GSC contains over a billion objects, with the majority having J and f magnitudes down to around magnitude 21. As the GSC was designed for pointing, multiple star systems are excluded to avoid confusion within the pointing system. The GSC is a popular catalogue, used in many astronomical packages. A digital version of the catalogue can be queried or downloaded from the Space Telescope Science Institute (STScI). The GSC divides the sky into regions and assigns a number to each source within a region. A third number (either a 1 or a 2) indicates which CD-ROM the star is located on and is now normally omitted. In the GSC format, α Ori is GSC 00129-01873.

The Sloan Digital Sky Survey

The Sloan Digital Sky Survey (SDSS) is a very large, very deep survey being conducted by the Sloan Foundation 2.5-m optical telescope at the Apache Point Observatory. The Sloan is very fast for a large telescope, being f/5 with 3° field of view. Sloan is equipped with a massive 120-megapixels CCD and a 640-fibre integrated Fibre Spectrograph. The SDSS contains over 500 million stars with magnitudes in u, g, r, i, z (i.e., Sloan filters) down to magnitude 22. The catalogue also includes 1 million spectra of galaxies. SDSS can be queried via the SDSS website http://www.sdss.org/.

2MASS is the 2 Microns All Sky Survey, an astronomical survey of the whole sky in the near infrared J, H, and K_s (s is for "short") utilising two 1.3 m telescopes located in Arizona and Chile. The survey contains 300 million stars or starlike objects and a further 1,000,000 galaxies or nebula-like structures. Completed in 2003, the survey contains objects as faint as approximately magnitude 15.

Although 2MASS is an infrared survey, it contains many optically red objects missed by other surveys, such as red dwarfs that are within the reach of many teaching instruments. Likewise, the catalogue has a high degree of completeness (i.e., most of the objects within its sensitivity limits are detected), and hence most non-solar system objects you are likely to observe will have 2MASS numbers. 2MASS catalogue numbers are a representation of an item's right ascension and declination. Hence, α Orion is known as 2MASS J05551028+0724255 in 2MASS.

The NGC and IC Catalogues

The New General Catalogue of Nebulae and Clusters of Stars (NGC) is a catalogue of nonstellar sources undertaken by the Danish-Irish astronomer John Louis Emil Dreyer in 1888. The catalogue contains a large number of galaxies that at the time were not fully understood, as well as star clusters and nebulae. In all, the NGC contains 7,840 objects, including all of those in the Messier Catalogue. NGC objects are numbered and prefixed with NGC; hence M31 is also known as NGC 224. The NGC catalogue was later supplemented by two further catalogues by Dreyer, known as the Index Catalogues (IC), which together add a further 5,386 objects, which are prefixed IC.

Uppsala General Catalogue

The Uppsala General Catalogue (UGC) is a northern hemisphere catalogue of bright galaxies down to magnitude 14.5, which should be within the reach of most northern teaching observatories equipped with a CCD. The UGC contains 2,921 galaxies, which are prefixed with UGC.

The Burnham, Aitken, and Washington Double Star Catalogues

The Burnham, Aitken, and Washington Double Star Catalogues are a generational series of double star catalogues initiated by Sherburne Wesley Burnham, who published his catalogue of 13,665 pairs of double stars. Burnham continued observing double stars, and that further work was included in the Aitken Double Star Catalogue (ADS), which extends the Burnham catalogue to 17,180 double stars, although like the Burnham catalogue, it is confined to mostly northern latitudes. Both Aitken and Burnham have been superseded by the Washington Double Star Catalogue, which contains 115,769 pairs of double star positions along with magnitudes and spectral types. Each Washington Double Star Catalogue object is denoted by its J2000 position and prefixed with WDS.

Henry Draper Catalogue

The Henry Draper Catalogue (HD) is a comprehensive catalogue of 359,083 (including various extensions) spectrally classified stars undertaken between 1886 and 1949 with a limiting magnitude of around 9, although in some cases it extends to as faint as magnitude 11. The spectral classification used by Henry Draper in the catalogue is the basis for the standard OBAFGKM spectrographic classification scheme used by astronomers and astrophysicists worldwide. Objects are numbered and prefixed HD, HDC, or HDEC depending on the catalogue version, although objects are numbered consecutively throughout all versions and supplements in order to avoid confusion. For example, α Orion is known as HD 39801 in the Henry Draper Catalogue.

General Catalogue of Variable Stars

The General Catalogue of Variable Stars (GCVS) and its sister catalogue, the New Catalogue of Suspected Variable Stars, are catalogues of, not surprisingly, variable, or suspected variable, stars. The GCVS lists the star's position and the class of variable (which the suspected catalogue omits). Stars within the GCVS are prefixed with GCVS, or in some cases, for example those in the SIMBAD database, they are prefixed by V*. The American Association of Variable Star Observers (AAVSO) also keeps a variable star catalogue, the International Variable Star Index, which currently holds over 20 million variable stars, many observed by amateurs.

6.4 Databases

Most of the catalogues listed above were originally supplied in printed form, which makes searching and cross-referencing somewhat tiresome. Later versions, such as the GSC and the SDSS, are supplied on CD-ROM (or DVD) that are machine readable and allow for fast searches. However, the trend is now to have online access with catalogues accessible by a web-based SQL search tool. In some cases, multiple catalogues are held at a single site and may be cross-matched with relative ease.

The most important online astronomical catalogue is **SIMBAD** (the Set of Identifications, Measurements, and Bibliography for Astronomical Data), an online catalogue formed from the concatenation of a large number of astronomical catalogues. At the time of writing, SIMBAD contained over five million objects, all located outside of the solar system. SIMBAD is searchable by both name and location and interacts with the Aladin application used for creating, amongst other things, finder charts. It supports multiple coordinate systems, including J2000, B1950, and Galactic, and it provides bibliographical references to celestial objects. SIMBAD can be accessed at http://simbad.u-strasbg.fr/simbad/ (Fig. 6.2).

Minor Planet Centre
For solar system objects (other than planets), the principal catalogue is the one held by the International Astronomical Union's Minor Planet Center (MPC). Like SIMBAD, the MPC is a searchable database containing over 600,000 asteroids, comets, moons, and other bodies, whose positions are regularly updated in order to provide accurate ephemeris information (the data that allows the calculation of an object's position in the sky). The MPC website has a useful New Object Ephemeris Generator, which you can use to help plan observations as well as for uploading solar system object data into planetarium software such as Stellarium. It also has a list of objects making their closest approach to Earth on the home page. The MPC's website can be found at https://minorplanetcenter.net/.

Exoplanets
Two invaluable websites for exoplanet observers is Exoplanets.Eu, which holds an online, open-source database of all confirmed exoplanets (http://exoplanet.eu), and the Exoplanet Transit Database (http://var2.astro.cz/ETD/), which has a vast database on transits, with viewable data as well as a prediction of upcoming transits and previous observational data on transits. You can also push your own data to the website.

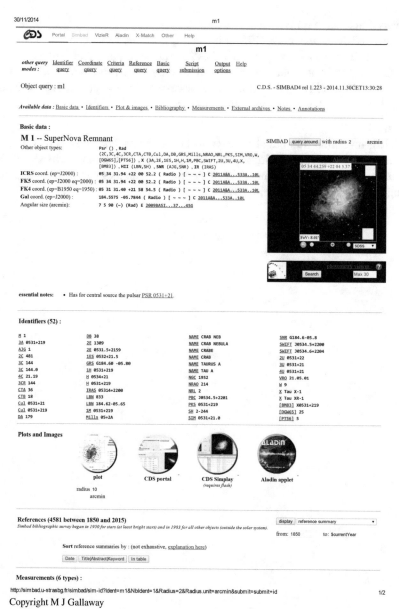

Fig. 6.2 Screen shot showing the results of a SIMBAD search for M1. The *top section* gives the position of the object in multiple coordinate systems, provides its angular size for resolved objects and its magnitude in multi-bands (although for some reason not in this case). A zoomable image of the object to aid with the location is on the *top right-hand* side. Below is a list of all the identifiers associated with this object, 52 in this case, and links to that catalogue. (Image supplied by NASA's Astrophysics Data System)

Infrared Science Archive

The Infrared Science Archive (IRSA) is a multiple infrared and submillimetre source and image database jointly operated by NASA and the Infrared Processing and Analysis Center (IPAC). It contains catalogues from surveys such as 2MASS, ISO, Planck, Spitzer, WISE, and Herschel. IRSA can be accessed via the web interface located at http://irsa.ipac.caltech.edu/applications/Radar/.

VizieR

VizieR is an online web-based astronomical catalogue service hosted by Centre de données astronomiques de Strasbourg (CDS), which also hosts SIMBAD. VizieR currently holds 12,488 astronomical catalogues. Catalogues can be cross-matched using the X-Match services also run by CDS.

Chapter 7
The Astronomical Detector

Abstract The charge-coupled device (CCD) is the foremost imaging device in observational astronomy. The CCD has high quantum efficiency and linearity compared to film, but it suffers from several sources of noise. We illustrate how to reduce noise by backing and applying flat, bias, and dark frames. We show how to determine the linearity of a CCD and how to recognize and deal with artifacts such as bad pixels, cosmic rays, and saturation.

7.1 The Detector

Before the introduction of photographic film, the only usable astronomical detector was the human eye. Even through an astronomical telescope, the vast majority of astronomical objects are not bright enough to be detected by the human eye, especially in colour. In fact, the **quantum efficiency (QE)**, the percentage of photons that trigger a physical response, is less than 5% for optical observations. This figure is for the light-sensitive rods within the retina; the colour-sensitive cones achieve much lower QE, and hence colour is seen in a very small number of astronomical objects when viewed with the eye. The cones are concentrated in the centre of the retina, and the rods around the periphery. A common technique used by amateur astronomers to see faint objects is to look at the target out of the corner of the eye in order to involve the light-sensitive rods. Another drawback to observations using the eye is that the exposure time of the eye is around 1/30 s, which is often an insufficient time for enough photons to hit the retina so that an image is transmitted to the brain.

The 1970s saw the introduction of the charge-coupled device (CDD) for professional astronomy, which is now a ubiquitous component of almost all cameras, astronomical or terrestrial. The QE of modern astronomical CCDs is now approaching 90%, and they are close to being nonlinear. The introduction of the CCD is one of the greatest driving factors in the major improvements in observation that have occurred in the last few decades.

CMOS (complementary metal-oxide-semiconductor) and CCD technologies were both developed at approximately the same time, the late 1960s and early 1970s, at

© Springer Nature Switzerland AG 2020
M. Gallaway, *An Introduction to Observational Astrophysics*,
Undergraduate Lecture Notes in Physics,
https://doi.org/10.1007/978-3-030-43551-6_7

Bell Labs. However, the fabrication of CCDs was considerably easier than that of CMOS chips, given the microprocessor manufacturing techniques used at the time. Consequently, CCDs received a considerable development advantage. Improvements in the production of very small scale circuitry and the large demand for consumer CMOS chips (most modern phone cameras are CMOS, as are most DLSR cameras) has resulted in the advantages between CCD and CMOS for astronomical imaging becoming increasingly similar. At the time of writing the second edition of this book, there was a gradual move by amateur astronomers away from CCD images towards CMOS, mostly because of the lower cost of CMOS compared to CCD cameras of the same size. Universities tend to spend more on imaging cameras but replace them less often, so it is likely, if you are using a university-based telescope or even a research instrument, that it will still be CCD based.

From a practical point of view, there is little difference between CCD and CMOS cameras. They use identical software and the same basic technology.

Both CMOS and CCD cameras use an array of light-sensitive cells, or **pixels**. You can consider each pixel to be a well with a series of balls (or in this case, electrons) around the rim. When a photon hits the well, an electron is knocked into it. At the end of the exposure, the number of electrons in each well is proportional to the number of photons that hit that pixel, with the number of hits each pixel received being known as its **count**.

Looking at the literature supplied with an astronomical camera or a manufacturer's website, you will see a large number of facts, functions, and figures associated with a camera. For example the frame transfer rate and whether it is backlit or frontlit. In general, this information will be at most interesting. However, there are a number of pieces of information you do need.

The **array size** is the size of the CCD or CMOS in pixels. The array size is not the physical size, as pixel sizes vary. In general, bigger is better. However, very large CCDs or CMOSs might not be fully illuminated by the telescope, so your camera should match your telescope.

The **pixel size** is the physical size of each pixel, which in most cases are square. Pixel sizes are typically in microns. You can use the array size and pixel size with the telescope's focal length to find the image scale, as discussed in Sect. 3.1.

The **quantum efficiency** (QE), as mentioned earlier, is the measurement of how effective the chip is at capturing photons. The QE is wavelength-specific and hence will change depending on the object being observed. For chips designed for optical observing, the peak QE is around 500–600 nm, with a very sharp dropoff in the blue end and a less sharp but significant dropoff in QE at the red end. Hence, you may find that some objects need longer exposures than others if their peak emission is outside the peak QE wavelength. Furthermore, QE is very sensitive to temperature, and you will not be able to achieve the expected sensitivity if the CCD or CMOS is running hotter than specified, one of the reasons astronomical detectors are cooled. Additionally, the published QE for a camera is unlikely to be achieved by every pixel in the array. This variation in QE can be a significant source of error, and it can be dealt with for the most part by a process known as flat fielding, which we will discuss at length later in this chapter.

The **ADU bit number** refers to the processing limitations of your camera. The ADU, or analogue to digital unit, effectively registers the impact of the electrons trapped in each pixel and converts the result into a number. The number of bits per pixel limits the maximum number the ADU can report. With an 8-bit camera, this limit is 256; for 16 bits, it is 65,536; and for 32 bits, it 4,294,967,296. Note that counts are dimensionless nonnegative integers. The **full well capacity** of a camera is the number of electrons each pixel can hold before it becomes unresponsive. A typical full well depth for a modern astronomical camera is 100,000–200,000 electrons. Having a higher full well capacity means that you can expose for longer times, which can improve the quality of the data within the image. However, most astronomical cameras have an ADU that cannot count as high as the full well capacity, and as a consequence, the number of electrons in each pixel has to be scaled. The scaling

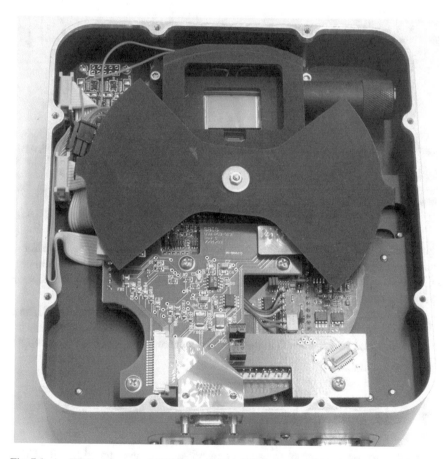

Fig. 7.1 An SBIG astronomical CCD camera. The CCD is clearly visible inside its window. The black tube on the side is full of desiccate to keep the air or inert gas dry. The I-beam-shaped metal object in the centre of the image is the shutter

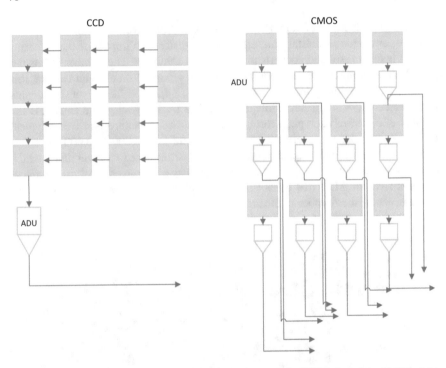

Fig. 7.2 This graphic shows the difference in reading between the CCD, left, and the CMOS, right. Note that each pixel in a CMOS has an ADU and a line to read the output, while a CCD has a single ADU. The ADU reads one pixel at a time down column zero and then shuffles the charge from column one to column two, repeating the process until all pixels are used. With the CMOS, the individual ADUs are queried consecutively

figure is known as the **gain**. You would be forgiven for assuming that gain for a camera is just the full well capacity divided by 2^{bits}, and in such a case, the gain of a 16-bit 100,000 capacity CCD or CMOS would be 1.5259. However, this is the lower gain limit, and typically a camera has higher gains than that (Fig. 7.1).

CMOS chips read differently from CCDs. In a CCD there is a single ADU, and all the pixel data passes through that ADU. In the case of a CMOS, each pixel has its own ADU. This results in much higher frame rates than those achieved with a CCD (Fig. 7.2).

7.2   Noise and Uncertainty in CCD and CMOS

We now move on to some of the sources of noise associated directly with CCDs. Noise in this context means anything that looks like part of the signal but does not come directly from the source. Noise and uncertainness are covered in general in

Chap. 11. This section deals largely with the noise sources directly linked to the CCD or CMOS, while this section discusses how to remove noise.

Not all pixels are created equal; there will always be a variation in the QE between individual pixels, which may change between exposures. This variation can usually be adequately addressed by the application of what is known as a flat field, which is discussed in detail in Practical 2. Knowing the flatness of your camera and the effectiveness of your flat field is essential in undertaking very sensitive photometry.

The electrons in each pixel are held in place by the application of a small charge. Intolerances in the makeup of the CCD or CMOS and impurities in the silicon from which it is constructed causes electrons to move between pixels, a process known as **charge diffusion**. Given the nature of the causes of charge diffusion, it will tend to be found in some parts of the CCD or CMOS and not others. It is also to some extent wavelength dependent. As with QE, flat fielding will deal in part with charge diffusion, and for most applications that you will encounter, it will not be a problem, as other sources of noise may well be dominant.

To fully understand the operation of the CCD, you will need to understand the processes that lead from an electron being liberated to an image being produced. For physical reasons, we cannot read all the pixels at the same time. Instead, the camera firmware and hardware undertake the following read process, as illustrated in Fig. 7.3. You can consider the CCD to be a matrix of pixels, with each pixel being identified by a row and column number. In general, the only pixel that is read is the one at 0, 0. Prior to being read, and as an additional precaution, a small number of additional electrons are added to the read pixel. This addition, known as the **pedestal**, is done to avoid a pixel going effectively negative (and potentially wrapping around, so that zero becomes a 65,535 count) during later operations. The ADU then converts the number of electrons into a count by applying the gain. This figure is then passed to an array. Once the 0, 0 pixel is read and the electrons removed from the pixel so that it now reads zero, the electrons in the next pixel up in the zero columns are transferred into the pixel at position 0, 0, and all the pixels in the column are moved down one. This process continues until the whole column is read. The camera then transfers the next column over to the now-empty first column, and all columns are moved over one. This cycle continues until all pixels have been read. The consequence of this is that **read noise** increases as we move towards the read column. Some very large CCDs are in actuality multiple CCDs combined, and therefore have multiple read edges. In this case, you will see the level of noise increase towards the edges of the array, as opposed to towards one edge.

The read process for CMOS is slightly different due to each pixel having its own ADU. In the case of a CMOS, there is no column transfer; each pixel reads out its value from its own ADU when requested, with the normal order being that the pixel at position 0, 0 reads out first. Although this process is faster than the readout of a CCD, it can suffer from a problem known as shutter blur when running at a high frame rate. Given that very high frame rates are very uncommon in astronomical imaging, shutter blur is unlikely to be an issue.

Most modern CCD cameras will enable you to perform a function known as on-chip **binning**. Binning is a process whereby neighbouring pixels are joined to

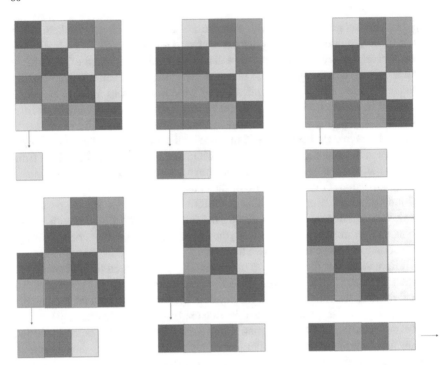

Fig. 7.3 Illustration of the read cycle of a CCD. An image is taken (*top left*), and the corner pixel is read. This pixel is then set to zero, and the charge of the pixel above it in the column is transferred to it. This is repeated until the whole column has moved down one pixel. The corner pixel is again read, and the cycle continues until the entire column has been read and is therefore empty. The adjoining column is then moved, and the next, etc. The read cycle continues until all pixels have been read

produce larger pixels. When performing this function, most cameras will average the counts for each of the bins, in which case, if we have 2 × 2 binning with four pixels responding with counts of 450, 445, 430, 424, the single pixel produced by binning will report a count of 437, the average value 437.25 having been rounded to an integer. Not all cameras do this averaging, and some will report the total value of 1,749, so be sure you know what binning process the camera uses before implementing it. On-chip binning happens before the data is downloaded. Binning not only reduces the number of pixels but also reduces the resolution of your final image. However, in general, most CCD/telescope combinations are overresolved, so that a small amount of binning does not affect the resolution. Notwithstanding the loss of resolution, on-chip binning has two major benefits. Firstly, it reduces download time, and secondly, it improves dark and bias noise, as they are averaged over multiple pixels.

Due to the integration of the ADU with the pixel, it should be noted that a CMOS-based camera is not capable of performing on-chip binning, and any binning that appears to be on-chip with CMOS has been done in software rather than in hardware.

The read process cannot be 100%, efficient and some loss during the transfer process can be expected. The rate of loss is known as the **charge transfer efficiency** (CTE). It affects the pixel diagonally opposite the read pixel the most and the read pixel the least. Consequently, the CTE results in a gradient appearing across the frame increasing outwards from the read pixel. For example, in a 512 by 512 array, the last pixel will go through 1,024 read cycles during a full-frame transfer. If the CTE is 0.999999, then the last pixel read will have lost about 0.1% of its electrons to CTE. The design of CMOS chips prevents them from being affected by CTE loss to the same extent as a CCD-based camera.

The process of reading the frame itself introduces errors. Using the read electronics in the system will have a tendency to add electrons to the system. This problem is exacerbated if you increase the transfer rate and also due to errors within the ADU itself. These errors, known as **readout noise**, are largely, but not entirely, dealt with by the subtraction of the bias calibration frame discussed in Sect. 7.5. Because CCDs use only one ADU, in contrast to CMOS chips, which use multiple ADUs, the read noise associated with the ADU is higher in CMOS-based cameras than in CCD-based ones because of variation in performance between the individual ADU processors in the CMOS. However, in general, ADU noise in both CCD- and CMOS-based cameras is low.

When we apply the gain to the electron count, we will in most cases get a noninteger number. However, as previously mentioned, the ADU is integer-only, so fractions are lost, either by a rounder processor or by just being truncated, so that at best, the ADU will report the count with an uncertainty of ± 1, unless of course, the gain is one. This is known as **digitisation noise**. Compared to other sources of noise, digitisation noise is not that significant. However, both the pixels and the ADU tend to have a nonlinear response as they approach their limit, just as photographic film becomes nonlinear. Nonlinearity is a major problem in imaging, as a pixel whose count is in the nonlinear range is effectively useless for most scientific uses. Nor is it immediately apparent whether a pixel is in the nonlinear range without first identifying the nonlinear range and determining whether the pixel lies within it. In Sect. 7.6 you will go through the process of identifying the nonlinear range for your CCD or CMOS. When exposing the CCD or CMOS to light, a number of thermal electrons leak into the pixels. These cannot be distinguished from those liberated by the interaction with a photon and can become a considerable component of the noise. This **dark current** is both temperature- and exposure-time-dependent and takes the form of a Poisson distribution, which we discuss in more detail in Chap. 5. The dark current should increase linearly with exposure time; hence if you are undertaking a long exposure of a faint object, the dark current may become very significant if left unaddressed. Likewise, an uncooled camera will have a significantly larger dark current than a cooled one. One of the most common reasons for bad image quality is either the failure to run the cooler or failing to deal with the dark current by applying a calibration frame, as discussed in detail in Sect. 7.5.

The last source of noise we will discuss that is directly, although not wholly, associated with the CCD or CMOS is flat field noise. In an ideal setup, the amount of light falling on the CCD or CMOS and the response to that light should be uniform, or flat. However, as we have discussed earlier, there are a number of sources of error that are structural in nature, such as variability in QE across the CCD or CMOS and the CTE in the case of a CCD. In addition, there is structural noise within the optics that contributes to the overall flat noise problem. These problems are addressed by **flat fielding**, which is discussed in Sect. 7.5. However, even the best flat field cannot entirely deal with this structural noise, and for some instruments, such as space telescopes, flat field noise can be the dominant noise source.

7.3   Binning and Subframing

An alternative to on-chip binning is off-chip binning (or post-download binning). Off-chip binning is a software action that is performed after download, so the benefits of reduced download time and the associated noise reduction are lost. However, the noise gain from averaging the noise over some pixels is still obtained, and for a CCD that does not support on-chip binning or for a CMOS based camera that cannot on-chip bin, off-chip binning is a suitable alternative.

When on-chip binning is performed, the calibration frames should have the same binning. Binning is not always a linear process, so you cannot just turn an unbinned calibration frame into a binned one using the software. Likewise, in performing off-chip binning, the same process must be applied to the calibration frames before subtracting them.

An alternative to binning is subframing. Subframing is a process whereby only part of the CCD is read. Subframing is a software-driven process and is not always supported. It involves taking an image of a target and selecting a region within the downloaded frame. Once that region is selected, only those pixels in the selected region are downloaded in future exposures. Hence with subframing, no resolution is lost, but you do not gain the noise reduction associated with binning. It is used mostly when download speed is needed, such as in focusing. The University of Hertfordshire's observatory uses a fast CCD camera to observe asteroidal occultations of stars. These are very transient events often only a few seconds in length. A typical astronomical CCD has a download time of a few seconds, so it is unsuitable for precision timing of such events. If we consider that the occulted star may fill only 100 pixels, by subframing, so that we download only the pixels with light from the target, we can significantly improve the download time and hence the cycle time of the camera. In fact, this camera has been run at a thousand frames per second for such observations.

A word of caution. Leaving the camera in subframing mode is one of the most common errors the author sees. It is not always entirely obvious to observers that they are taking subframed images. So make sure you turn subframing off when you have finished using the camera or when you are no longer subframing.

CMOS Versus CCD

As can been seen from the above section, there are pros and cons to both CCD and CMOS chips. In general, research-grade CCD chips still outperform CMOS-based cameras for astronomy. They are mostly used for **stare mode** observations, that is, long-exposure observations at a single unmoving object of constant brightness. They have lower bias and dark noise current than CMOS chips and are capable of on-chip binning, unlike CMOS-based cameras. However, they suffer from CTE, although CMOS cameras can suffer from **edge glow** due to the large amount of electrons needed to power the individual pixels on a CMOS chip. Hence, CMOS pixels are considered active, while there is no power to a CCD pixel when it is exposed, and so it is considered inactive. CMOS pixels are therefore considerably more complex than the pixels within a CCD, and therefore dead and bad pixels are more common and are likely to multiply faster as the camera ages for reasons other than that there are more things to go wrong with each pixel.

A CMOS camera, however, is generally less expensive than a CCD-based camera, permitting you to get a larger sensor for the same money. CMOS cameras have much higher frame rates that CCD, meaning that they are more suitable for lucky imaging than CCD-based cameras.

At the time of writing in mid 2019, Sony had decided to stop the development of their range of CCD chips and eventually will be closing their production line down. Kodak, which makes the KEF range of CCD chips, have yet to announce any change, but it is possible that as demand for CCD chips shrinks and the performance difference between CCD and CMOS diminishes, CCD production may end for all but a few small manufacturers making CCD chips for very specialist markets.

7.4 The FITS File Format

The Flexible Image Transport System (**FITS**) is the standard format for astronomical images and image-derived data such as spectra, and files containing such data may have either the .fit, .fits, or .fts extension to the file name. The FITS file format is designed to carry metadata (data about the data), two-dimensional image data, and three-dimensional data such as radio and millimetre data. The **FITS header** holds the metadata in a series of labels known as **cards**. Each card will have a label stating what it is, for example EXPTIME for exposure time, and a piece of data, for example 10. This card tells the user that this was a 10 s exposure, but it also tells your software that it was a 10 s exposure. You can add and change header cards using any number of astronomical packages. Just keep in mind that if you use something that is a standard card name, it might cause problems later. Robotic telescopes, as well as taking the image, run the data through a a series of operations on the image, called a **pipeline**. All these operations are noted by the insertion of a card, or cards, in the header. The addition of these cards allows the user to identify what has happened to the image and allows the processes to be reversed if needed (which isn't always possible).

The FITS header may also contain information about the translation between pixel location and astronomical coordinate systems such as Galactic. The translational system is known as the World Coordinate System (**WCS**). If the application you are using has the ability to display pixel WCS position but does not do so for the image you are displaying, the WCS has likely not been calculated for that image. This is easy to resolve as we will show later, but it is not always necessary.

Unlike other image formats such as JPEG and PNG, FITS files are not compressed, so they tend to be quite large; however, they contain all the data taken by the camera, unlike more traditional imaging formats. If you are familiar with digital photography, you may have noticed that your DSLR camera can save images in RAW format; when you select RAW, the number of images you can store on your memory card is much reduced. This is because RAW is the terrestrial version of FITS. Most professional photographers will take images in RAW, as they can then be easily manipulated. For the same reason, professional astronomers use FITS, as should you. If you need to make a JPEG image from a FITS, it is fairly straightforward, and most FITS display applications have a "Save As" option. You should, however, always keep your scientific data in FITS. FITS is supported by a wide range of computer languages with libraries available in C, C++, Perl, R, and Python, as well as many others. Using code to manipulate FITS files is very straightforward and very powerful.

7.5   Calibration Frames

Read noise can be easily addressed by taking a **bias frame**. A bias frame is an exposure of zero seconds, with the shutter closed. Hence, it forces the camera to go through a read cycle while not exposing it to any light. Therefore, any electrons registered have not come from photoelectric interaction in the CCD, but are due to random fluctuations introduced by the reading process, electrical components in and near the camera, and to a very small extent cosmic rays (see Sect. 7.7). When you take an image of your target, a **light frame**, the bias frame should be subtracted from your light frame. This subtraction process normally happens automatically with modern camera control software. When using a previously generated bias frame (and there is nothing wrong with doing this, as the variation of the bias over time is very small), it has to be taken at the same camera temperature as the light frame. Subtracting the bias frame removes any read noise, as well as the pedestal that was added during the reading process. Although pixel values within a bias may be quite high, depending on the camera and the software you are using, the standard deviation across the array should be very low, perhaps only a few counts. If it is not, then you might have a reading problem or a faulty piece of equipment, for example a bad camera cooling fan. In general, a good bias frame is the median of a very large number of bias frames. Taking the median reduces the amount of random noise associated with individual frames and to some extent addresses cosmic rays.

With so many pixels on a modern camera—the camera on the Gaia space telescope has over a billion—clearly there will be some random noise in the sensor. To address this, we take a **dark frame**. A dark frame is an exposure, with the shutter closed, of a duration equal to the light frame. The dark frame, as with the bias frame, is subtracted from the light frame to remove any inherent problem with individual pixels. Dark frames should be fairly linear, i.e., if you double the exposure time, you should double the count within the pixel. You also should by now have noticed that a dark frame must also include a bias frame, so in general, the removal of the dark also removes the bias, but not always. Again it depends on the software you are using. You should also notice that a dark must be linear only after the nonlinear bias has been removed. Again, as with bias frames, dark frames should be taken at the same temperature as the light frame. Typically, they are taken either immediately after or before the light frame is taken, and in many cases, the software you are using

Not all pixels are created equal. Even on the best science-grade array, some are more sensitive, some less so. The hole in the secondary mirror of a Cassegrain telescope, the secondary, dust on the filters, the correcting plate, or the CCD window and nonuniform illumination of the chip due to the telescope aperture being circular and the array being rectangular all need to be accounted for. This is done by taking a **flat frame**, which, unlike biases and darks, are used to divide the light frames. Flat frames require that the camera be exposed to a uniform level of light until the median pixel count reaches a set limit (for most 16-bit cameras this will be around 30,000 counts). This has to be done for every filter mounted on the camera. Normally, multiple flat fields will be taken, and their median value for each pixel used to create a science flat. Typically, flats are taken during twilight using a position opposite the Sun, as that is considered the flattest (i.e., most uniform) part of the sky, with multiple flats taken and their mean value used as the master flat for that filter. Flat frames taken in this manner are known as **sky flats**. Very good flats are extremely challenging to achieve, and because they account for the contamination of the optics, they need to be updated much more often than dark or bias frames. Hence, there is now a tendency to do **dome flats**. Dome flats use the illuminated inside of the telescope dome as their light source, rather than the sky. Dome flats may be done at any time and are more consistent, because the environment in which they are taken can be controlled. However, they do not cover the entire optical pathway (because they don't include the atmosphere), and they may not be as uniform as sky flats. There is currently considerable debate about sky flats versus dome flats, the methods used to produce skies and domes, and their impact on many aspects of observational astronomy, as will be discussed later.

When an image of an object is taken, it is known as a **light frame**. You will need to subtract the bias and dark frames from your light frame and divide the result by your flat frame (minus its bias and dark component) to create a **science frame**. Many professional astronomical telescopes will do this as part of the **pipeline**, a series of computer programs that together turn an astronomical image into a usable science image (Figs. 7.4, 7.5, and 7.6).

Fig. 7.4 Bias frame from an SBIG 1301 CCD. The *bright edge* on the left-hand side is the noise generated by the read process. (Image University of Hertfordshire)

Fig. 7.5 Dark frame from an SBIG 1301 CCD. The *bright corner* on the left-hand side is the noise generated by the read process. The bright points scattered randomly across the frame are hot pixels, as discussed below. (Image University of Hertfordshire)

Super Flats and Fringes

In some cases, it is necessary to take more advanced calibration frames, for example when it is impossible to do a dome flat or when very high precision is needed. In such a case, a **super flat** may be required.

In simplest terms, a super flat is a normalised flat frame constructed by taking the mean of several on sky images that contain very few stars (no stars at all would be pre-

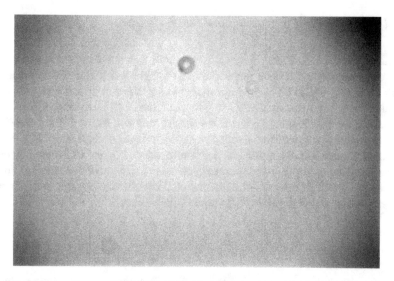

Fig. 7.6 Flat frame from an SBIG 1301 CCD. The gradient across the frame is partly caused by read noise but also by the chip not being uniformly lit. The *dark circles* are out of focus dust in the optical path. The *very dark circle* is dust on the CCD window. (Image University of Hertfordshire)

ferred, but such is effectively impossible) and no extended objects such as galaxies or nebulae. The stars in the field can be removed using a sigma clipping algorithm that removes pixels with values outside a set standard deviation and replaces such pixels with either the mean value of the image or by the average of the eight pixels surrounding the pixels being clipped. super flats, as the name suggests, are very flat and should be flatter than both dome or sky flats. However, modern dome small telescope flat fielding techniques using electroluminescence have become very sophisticated and may be very close to super flats in quality.

Fringes are an unwanted interference pattern caused by long-wavelength photons passing through the thinned substrate of a back-illuminated CCD. The pattern has been likened to waves on water. You can think of a backlit CCD as the reverse of front-lit. Whereas in front-lit CCDs, the wiring for the CCD is on the exposed surface of the chip, for backlit CCDs it is at the rear, and the light has to pass through the substrate on which the CCD is mounted in order to be detected. Backlit CCDs have some advantages such as being cheaper to manufacture and being more sensitive to light. More modern backlit CCDs have reduced fringing, and it is unlikely that you will need to address this problem (although applying a super flat is one way of doing so).

7.5.1   Practical 1: The Bias and Dark Frames

In this, the first of the practical sessions, you will learn how to take dark and bias frames. This practical can be undertaken in daylight and in poor weather, as the dome does not need to be opened nor the lens cap removed. It will give you a first introduction to your camera control software and the FITS Liberator package.

Leaving the bias and read noise aside for the moment, as we take progressively longer exposures even with the shutter closed, we find a mysterious build-up of counts within the pixels. These are thermal electrons produced by the CCD itself entering the pixel and are one of the main sources of errors in CCD-based observations. In this practical, you will investigate both the dark and bias frames in order to understand their impact on uncertainties in CCD-based observations.

Taking the Frames

1. Set up the telescope, ensuring that no light is entering the camera by placing the lens cap on the telescope. The telescope does not need to be powered up and does not need to be tracking. Connect it to the camera using your camera control software.
2. Using the camera control software, make sure you have the binning set to 1×1 and that you have subframing turned off. Turn off the cooler and let the CCD warm up to ambient temperature. If your control software has a warmup option, you may use this to save time.
3. Set the camera temperature just under ambient, so if the uncooled camera is at $12\,°C$ set it to $10\,°C$. If your camera control software has the ability to select between frame types, then select bias. If your software can do autocalibration, make sure this is also turned off. Once your CCD is down to temperature and is stable at that temperature, take five zero-second exposures, saving each frame. It would help if you saved each frame with the temperature as part of the name. Your camera control software should put the temperature in the header, but it might not.
4. Once you have taken your first set of bias frames, reduce the temperature by five degrees. Again wait until the temperature is stable and then take a further five frames. Continue this process until the camera reaches either its normal operating temperature or the point where the cooler is running at 100% in order to maintain the temperature.
5. You will now need to take your dark frames. Warm the CCD up to ambient temperature, and as before, cool it down to just a few degrees below ambient. If you have the ability to set the software to dark frame, do so.
6. Once the temperature is stable take five dark frames with an exposure time of 1 s.
7. Increases the exposure time to 5 s and take and save a further five darks. Continue this process for 10, 30, 60, 120, and 240 s. Once this is done, cool the camera

down by 5 °C and repeat the cycle. Take 1, 5, 10, 30, 60, 120, and 240 s exposures at the new temperature.

8. Continue this cycle down to the same temperature at which you took your last bias frame.

9. Once you have downloaded all your frames, open each frame in FITS Liberator (Fig. 7.7). The image statistics panel on the right-hand side, next to the image, shows the Image Statistics. Record the Median and the StdDev from the Input column for each image. This represents the median pixel value in the array and the standard derivation of the array, i.e., how much it varies. Do this for all your images and record the results on spreadsheets as two tables, with one table for the bias frames and one for the darks. You can use the Image Header Tab in FITS Liberator to confirm the temperature at which each image was taken and the exposure time.

10. For each block of five images, for both the bias and dark frames take the mean of the median count values and of the temperature, making a note of the spread of the median count values (the maximum value minus the minimum value), as this will be used for your uncertainties. Use these to make two new tables, one for the bias frames and one for the dark frames.

11. Looking at the bias, plot the mean bias median value (y) against mean temperature (x). Your error for each point should be

$$\pm \frac{\Delta x}{2\sqrt{N}}, \tag{7.1}$$

where Δx is the spread of that point, and N is the number of measurements in that point, i.e., in this case, five.

12. Think about what the point at which the line crosses the y-axis represents and whether the temperature has a large impact on the bias. From your standard deviation data, does the bias frame change much across the CCD? If it does, what causes this?

13. Moving on to the dark frames. For each exposure time, plot mean bias median value (y) against mean temperature (x) using (7.1) for your error bars. Do this for each exposure time, plotting all seven exposure times on the same graph if possible.

14. Keep in mind that the bias is part of the reading process, and so it appears in the dark frames. You will need to address this in your analysis.

15. If you have access to another camera, particularly a different model of camera, and have the time, repeat the process.

Analysis

You should write your report up in standard laboratory style: introduction, method, results, and conclusions. You should include all your tables and your plots as well

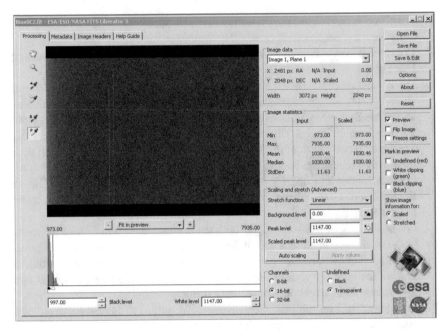

Fig. 7.7 The above figure illustrates a bias frame loaded into FITS Liberator. The Median and StdDev figures required are highlighted

as notes about the camera and software used in addition to any problems you may have encountered. Your report should include the following:

1. A description of the shape of your curves and an interpretation of how the curves vary with exposure time and temperature?
2. What was your bias level? How does the bias vary with temperature and internally across the CCD?
3. How does the dark current vary with the exposure time and temperature? (Don't forget that there is a bias in your dark, so you need to subtract the effects!) How does the dark frame variation across the CCD compare to that of the bias frame?
4. How significant is the dark current when compared to the bias current, and how might we deal with both the additional counts caused by the bias and dark and their associated uncertainties?

7.5.2   Practical 2: Flat Frames

In the second practical you will be taking flat frames. Flat frames are a vital part of the calibration process and are the calibration frame type that needs to be updated the most often. As you will see, they are also a primary contributor to uncertainties. In this practical you will familiarise yourself with the flat field method, the operation of the camera, and the use of SAO DS9. The analysis will give you an understanding of both the importance of flat fielding and the impact that poor or irregular flat fielding has on image quality.

Flat Fielding

1. Turn the dome flat field lights on. These might just be the normal white lights, but do not use the dome red lights if the dome has them, as you want the light going into the telescope to be as spectrally as flat as possible. Take the lens cap off the telescope.
2. Rotate the dome so that the flat field board is facing the telescope. You may have to move the telescope in alt to get it directly facing the flat field board. Rather than power the telescope down, turn its tracking off. If your telescope has a flat field diffuser, put this in place.
3. Power on the camera, connect the control software, put the cooler on, and set it to the correct temperature. While you are waiting for the camera to get down to operating temperature, make sure the binning is either off or set to 1×1 and that the subframing and any autocalibration are off.
4. Change the filter to R or $\acute{r}$ and take a short exposure. If your camera control software has the option to select the frame type as "flat," select it.
5. Take a very short exposure, not more than 1 s. Once it is displayed on the camera control software, check to see the mean count near the centre. In MaximDL this is done by bringing up the information window (ctrl-i) and selecting Aperture from the Mode drop-down. If you are not using MaximDL, it is likely that your software has a similar capability. Failing that, you could open the image in FITS Liberator and use the mean value in the image statistic box. You are trying to obtain a mean count of around 30,000–40,000 for a 16-bit camera. If you are getting too many counts, you should reduce the exposure time until it is in the correct range. Likewise, if the count is too low, increase the exposure time.
6. Once you are happy with your exposure times, take five flat fields and save each one. Don't forget to record your exposure time in your lab book.
   You can perform the analysis of your flat field images in DS9. You will need to determine how flat your field is. To achieve this, you need to measure the average count in a number of small regions spread over the field. This cannot be done with FITS Liberator, whence the need to use DS9.
7. Open your first flat in DS9 and set the scale. Using the Scale menu, select 99.5%, and using the Zoom drop-down menu, select zoom to fit the frame.
8. You will now need to set up your regions. From the Region menu, select Shape and then Box. Double click on the centre of the image, and a box will appear. Single click on the box and drag it out so that it is square with width about an eighth the width of the frame.
9. Move the box to the top left-hand corner of the frame. From the Edit menu, select Copy and then paste (or press ctrl-C and then ctrl-V). This is creating

a second box colocated with the first. Drag this box to the bottom left corner. Repeat until you have something like Fig. 7.8.

10. From the Region menu, select Select All (or press ctrl-A). This selects all the regions. Now again from the Region menu, select Save Regions. Enter a file name and then select DS9 for the format and Physical for the coordinate system. This will save your region layout for future frames.

11. Pick a region and click on it. This will bring up some of the information on the region including its centre. Record the centre, making sure it is in physical units, not, for example, RA and Dec. Use trigonometry to determine its distance from the centre of the array. (Hint: you can find the array size in the header.) With the region information box, open Select Apply and a second box with more information on the region. Find the mean count and record it on a spreadsheet along with the region's distance from the centre.

12. Repeat this for the remaining four frames, although this time you can load your saved regions in, which is quite a time saver.

13. Plot distance against mean count on a spreadsheet for each frame on the same graph.

Analysis

Write your report up in standard laboratory style: introduction, method, results, and conclusions. You should include all your tables and your plots as well as notes about the camera and software used in addition to any problems you may have encountered. Your report should include the following:

1. A description of the overall appearance of your flat field image, including the small-scale and large-scale uniformity. Are there any unusual features in the flat field, and if so, what do you think they are caused by?

2. Your flat fields include a bias and a dark current. Given what you know about these two calibration frames, how do the errors they introduce compare to the flat field?

3. There might be variation between the edge of the flat field and the centre along each of the axes. If there is, can you explain it? What would the impact be if we imaged a star in the centre of the field and then at an edge? Can you quantify this effect?

4. When the observatory staff take flats, they take multiple copies and combines them into a master flat. How do you think this is done? Are they just added?

5. Dark frames are temperature and binning sensitive. Are light frames also temperature and binning sensitive? How do you think the choice of filter affects the flat? Do you think we need to flat field for every filter, or will just one do?

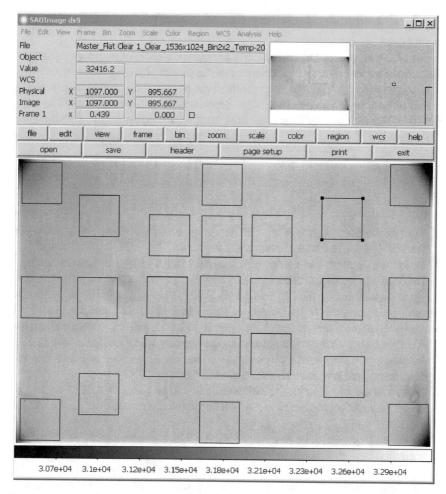

Fig. 7.8 Example of the DS9 region layout for the flat field practical

7.6 Linearity of CCD

Unlike photographic film, which does not respond in a linear manner when exposed to light, the response of a CCD is linear over most of its dynamic range (around 1000–65,000 counts for 16-bit cameras). Hence, doubling the exposure time doubles the count in an exposed pixel. However, when the counts begin to get high, such as with very long exposures or very bright objects, CCDs tend to become progressively nonlinear before becoming saturated and no longer responsive.

It is important to determine where this point for nonlinearity lies so it can be avoided. This is normally determined using a modified version of the flat field technique outlined in Practical 2.

7.6.1   Practical 3: Measuring CCD Linearity

In this practical, you will be investigating how the properties of telescope design and the CCD detector can affect the accuracy of the results you obtain from your astronomical imaging observations. This practical can be undertaken either with the dome closed or by imaging a star. The choice depends on weather conditions. The point of loss of linearity of a CCD is one of the most import components of photometry, the measuring of a star's brightness, and it is important that you understand why it happens, where it happens, and how to avoid it.

Taking the Data

1. Turn on the dome lights. If you can dim them, turn them down enough that you can still just see the control computer's keyboard.
2. Rotate the dome so that the flat field board (or if there isn't one just the inside of the dome) is facing the telescope. You may have to move the telescope in Alt to get it directly facing the flat field board. Rather than power the telescope down, turn its tracking off. If your telescope has a flat field diffuser, put it in place.
3. Power on the camera, connect the control software, put the cooler on, and set the temperature. While you are waiting for the camera to get down to operating temperature, make sure the binning is either off or set to 1×1 and that the subframing and any autocalibration are off.
4. Select the narrowest filter. In most cases, this will be either OIII, Hydrogen α, or an SII Filter. If you do not have any of these in the filter wheel, use an R or B (but not a V).
5. Take a 1 s exposure, save it, and open it in DS9. Scale the image to 99.5%. Create a circular region. It needs to be fairly small, say 100 pixels in area, and locate it near the centre of the field. You will need to save the region so you can load it into future frames. Look at the mean count within that region. If it is over 30,000, your exposure is too long or your dome lights too bright. Try to reduce them until you have 5,000 counts or less.
6. Take a succession of light frames, increasing the exposure time with each. I suggest 2, 5, 10, 15, 30, 60, 90, 120, 180, and 240 s exposures. After you save each one, open it in DS9, load the region you saved in step 5, and make a note of the mean count. You can stop when the mean count hits about 64,000 (for a 16-bit camera), at which point the CCD is, of course, saturated.
7. Plot the mean count against exposure time. You may wish to try fitting a line to your data. If you have undertaken practicals 1 and 2, you will have some idea of the size of your error bars, but remember that you are plotting the mean count, not the actual count.

Analysis

As previously, you need to write your report up in standard laboratory style. You should include your table of exposure times, mean counts, and your plot as well as the standard notes about the camera and software used in addition to any problems you may have encountered. In particular, your report should include the following:

1. In what circumstances is the detector response of a CCD preferred to that of photographic film?
2. Describe the shape of your plot, in particular how the gradient changes with exposure time if it does.
3. Try to explain the significance of any change in the gradient and its impact on performing precision measurements such as those needed to detect exoplanets using the transit technique.
4. Your plot may not be straight through the whole length. Can you suggest a process that might cause this effect?

7.7   Bad Pixels and Cosmic Rays

It is unlikely that any astronomical camera will have every pixel working, and more pixels will be lost as the instrument ages. The most common fault is a dead pixel, a pixel that will either not read or is no longer sensitive to light. Dead pixels will read as having either zero counts or a count near the bias level and are difficult to see in an image. Hot pixels are pixels that read artificially high, either at or close to the maximum value (65,536 for a 16-bit camera) and appear as bright isolated points in the image. You may also see dead columns. These can be caused by a faulty pixel preventing the read process in that column. These faults may cause problems when you try to extract data from the image, so you need to be aware of them. Most astronomical imaging software has a function that attempts to deal with hot and cold pixels, although no process, besides physically avoiding that part of the CCD, is totally satisfactory. The most common method works by identifying pixels with values under or over a set value and replacing the faulty pixel with the mean or median value of its neighbours.

A CCD cannot distinguish between photons and other high-energy particles such as those emitted naturally by radioactive decay from the surrounding environment. Given that the CCD is effectively shielded, local events rarely have enough energy to penetrate the casing and register. However, cosmic rays, very high energy particles, many thought to be from supernova explosions, stream into the Earth's atmosphere, causing cascades of high-energy particles that pass through the Cassegrain aperture of the mirror and dump their energy into the CCD. This causes a trail of high registering pixels that appear hot when read, which of course they are not. Cosmic ray detections on CCD images are difficult to remove using standard methods, as they appear at random and tend to leave a jagged trail. This means that the "median of the neighbouring pixels" process above will not produce a satisfactory result.

7.7.1   Practical 4: Bad Pixels and Cosmic Rays

In this practical, you will use the CCD as a cosmic ray detector. In order for this to be effective, you need to remove the hot pixels that appear to result from cosmic rays. This practical is written for Diffraction Ltd.'s MaximDL package, but you could use any suitable package.

Measuring Cosmic Rays

1. With the dome closed and the lens cap on, connect your control software to your camera.
2. Ensuring that the camera is down to temperature, that the binning is set to 1×1, and there is no subframing, take two 15 min dark frames and save them as Frames 1 and 2.
3. Subtract one frame from the other.
4. Load the resulting image into DS9 and identify the cosmic rays in the image by putting a region on them.
5. Using the hot pixel tool, find the number of pixels affected by cosmic rays.
6. Do this for a number of different exposure times and plot cosmic ray count against dark exposure time.

Analysis

Your analysis should include the following:

1. Your images, including the images with the cosmic rays marked. Also include the results of the hot pixel tool and your plot.
2. How do cosmic rays differ from hot pixels?
3. Given the size of the array, proportionally how many pixels are affected?
4. How does the exposure time affect the number of cosmic rays? Why do you think this is?
5. In undertaking precision photometry, the appearance of cosmic rays has to be addressed. Given that dark frame subtraction is not effective, can you suggest another way of dealing with cosmic ray strikes on CCD.

7.8 Gain

As stated previously, each pixel on a CCD is, in effect, a small capacitor storing electrons liberated by the effect of an incident photon. The value that is actually read is the number of electrons within the well. However, what is reported is the count. A pixel can often hold in excess of 100,000 electrons, which is more than a 16-bit analogue-to-digital converter (ADC) can report. Hence, the number of electrons is scaled. This scaling factor is the gain G. Your observatory staff should be able to tell you the gain of the camera you are using, and it should also be recorded in your lab book and your images as part of the FITS header.

As always, no two cameras are created equal, and there may well be a slight differential between the specified gain and the actual gain. In an ideal world, the gain should be the maximum well capacity in electrons divided by the maximum number the ADC can hold. In reality, the gain is seldom that high. In the next practical we will determine the gain of your camera.

7.8.1 Practical 5: Measuring Camera Gain

In this practical you will find the gain of your camera. You will need the camera gain when calculating uncertainty in your observations. If new, the camera may very well be operating near the manufacturer's published level, but with age it may move away from that value. You will be using the skills you have developed during the previous practicals in this book.

Taking and Reducing the Gain Images

1. With the camera down to temperature and the lens cap on the telescope, take a series of darks with 1, 2, 3, 4, and 5 s exposures. Take two darks for each exposure time. Ensure that the binning is set to 1×1 and subframing is off. If you have the dark frame option on your control software, turn it on.
2. Take the lens cap off and put the dome lights on. Set the telescope up for taking flat frames as previously.
3. Take pairs of flat frames for 1, 2, 3, 4, and 5 s exposures. Start with the 5 s exposures. Using FITS Liberator (or another appropriate package), ensure that you are not entering the nonlinear region of the CCD, hence the need to start with the long exposures. If you are entering the nonlinear region, swap to a different filter and try again.
4. Using an appropriate package (for example MaxImDL or a Python script), generate a mean image of each of the pairs of dark frames and subtract these from the flats of the same exposure time.

5. Taking each pair of images, subtract one of the pair from the other to make a variance image.
6. For each of your five variance images, open them in FITS Liberator and find the mean value and the sigma value (σ).
7. Plot (in Excel, for example) $\sigma^2/2$ against the mean and then fit a straight line to your points.

Analysis

When writing up your report in the standard format, you should include the following:

1. What feature of your plot do you think indicates the gain? How does it compare to the gain reported by the manufacturer?
2. How does the measured gain compare to the maximum gain?
3. As you subtract the pairs of flats to create your variance images, do you think it makes a significant difference which is subtracted from the other? Why?

7.9   Saturation, Blooming, and Other Effects

Each pixel can take a limited number of electrons before it becomes full. This is known as **saturation** and should be avoided as much as possible. When a saturated pixel is read, the reading process causes electrons to flow into the pixels above and below the saturated pixel, often more in one direction than the other. In extreme cases, this causes a spike to appear across your star, a feature known as **blooming**. Once an image has a bloom in it, the area around the affected part of the image is effectively a loss as far as science is concerned. Some cameras have anti-blooming gates, which act as capacitors between the pixels, and thereby stop stray electrons. In general, for scientific applications, the disadvantages of cameras with anti-blooming outweigh the advantages. If you have a camera with this feature and you intend to take scientifically useful frames (as opposed to just nice images), then turn the anti-blooming off if you can and keep the maximum pixel value well below the saturation point. In press images from large telescopes, you often see stars with crosses and halos around them. These are **diffraction halos** and **diffraction spikes** caused by light being diffracted around the secondary and the secondary supports. You should probably avoid these becoming too bright when doing science frames, as they will affect your measurements. You should also be aware that these are sometimes added after the image has been processed for artistic reasons.

Another problem that may be encountered with long exposures is **residual bulk image (RBI)**, another effect of saturation. If a pixel becomes saturated, further photon strikes can cause electrons to move out of the pixel and into the substrate on which the CCD is mounted. When you move to another target, these electrons leak back into the pixel from whence they originated and create a ghost image of the previously observed star. Some cameras have an IR flash that can help prevent RBI by prefilling the pixels. For cameras without this feature, removal of RBI is troublesome, as it persists for some time (the author has seen RBI last 20 min). One way to fix this is to warm up the camera and allow thermal electrons to displace the RBI electrons, although the best way to fix RBI is, of course, to not let it happen in the first place by not saturating.

On occasion, when taking an image you may see a strange, almost ray-like feature vignetted across the image. A very bright star, planet, or the Moon located just outside the field of view of the telescope but close enough that light is still entering the telescope is the usual cause. In general, there is no alternative solution for this other than not imaging under those conditions. You should not confuse this problem with dome clipping, whereby the edge of the dome is very near, or actually in, the field of view of the telescope and stray light from inside the dome is reflecting into the aperture. Nor should you confuse these with very straight single or double lines crossing your image. These are satellite and aircraft trails.

You might also see fractal-like structures appearing at the edges of an image. These are ice crystals forming on the window of the CCD. Within your astronomical camera, the CCD is sealed within an airtight windowed box. The box is normally filled with an inert and very dry gas such as argon or nitrogen. Over time, water vapour gets into the window and can freeze on the inside. If this is happening, warm the camera up again, and then cool it slowly in steps. This should stop the icing, but inform the observatory staff, as they will want to address this issue, as it might lead to CCD failure.

7.10 DLSR Astrophotography

If you do not have access to an astronomical camera, then it is likely you will be using a digital single lens reflex (DSLR) camera. These cameras are the standard for both professional and serious amateur photographers for terrestrial photography. They will have good quality CMOS sensors, good sensitivity to light, removable lenses, and relatively low noise compared to compact cameras and phones. Some models now have a live view mode that displays what the sensor is seeing on the screen. There is a new generation of mirrorless DLSRs coming onto the market that allow the user to effectively see the sensor's output. As they are so new, it is difficult to gauge how mirrorless DLSRs will perform in astronomy. However, they have a reputation for having short battery lives, which given the long exposure times used in astronomical imaging is disheartening.

A DSLR camera provides excellent low-cost astronomical images. You need to mount it to your telescope using a T-piece, which replaces the lens of the camera. Just as a camera lens is specific to a camera manufacturer, so are T-pieces. So if you have a Nikon DLSR, you need a Nikon T-piece. You will also need a remote shutter release, most of which are now infrared or wireless. Some DLSRs have camera control software bundled with them (or available as an add-on). If your camera has this feature, it is worth having, as it means you can control the camera without touching it, thereby avoiding moving the telescope in the process.

In standard, nonastronomical, photography, there is what is known as the **exposure triangle**, which consists of the ISO, the aperture, and the exposure time. In the case of astronomical photography, you have a fixed aperture, so that cannot be changed. Most photographers assume that the ISO is the same as the old film speed and represents the sensitivity of the sensor. Such is not the case. Changing the ISO on a DSLR just changes the preamplification from each pixel, which improves low light performance at the cost of increased noise. Obviously, this is not a desirable side effect. As a result, you should keep the ISO of your camera set to its base level, which is normally 100 or 200. You will not be using white balance, so either turn that off or set it to daylight. If your camera has an exposure compensation setting, that should be set to zero. The camera should be set to manual mode. Every DSLR camera has a maximum exposure time; the author's own camera is limited to 30 s. For exposures beyond this, you need to set the camera to *bulb* mode. In bulb mode, the shutter stays open for as long as required. It is not determined by the camera.

There are a number of problems with using a DSLR rather than an astronomical camera. The first is that commercial DSLRs are much noisier than astronomical cameras, which are built to be cooled and to have low noise. DSLRs are also non-photometric, due to the fact that each pixel is filtered by the manufacturer. Lastly, the processor controlling a DSLR may perform some image processing such as dark and bias removal before the image is downloaded. It might be possible to turn this off. If you have a noise reduction option, this should be turned off, since although it might be reducing the noise, you do not know how or what else it is removing.

Lastly, if you are using a DSLR, ensure that it is saving the files as RAW, which might also be called NEF or ORF depending on the manufacturer. This produces a file that is uncompressed and has all the image data in it. RAW files are large, especially for large sensor cameras, and they may seem very large compared to JPEG, but you should always use RAW.

7.11 File Formats

When a digital image is downloaded from an imaging device, the data is stored in a file format. The file format describes how the data is encoded and how it may be converted into an image on a screen. Broadly speaking, there are three file formats types. Uncompressed formats such as FITS store data as a matrix, with each data point (or several data points) in the matrix representing a pixel in the image. Uncompressed file formats tend to result in large file sizes. A 640×480 pixel image in such a format would result in a 900 KB file size.

Compressed formats use a compression algorithm to reduce the file size, and a single data point may represent many pixels in the image. The compressed format may be lossless or lossy. A lossless compression format can restore an image to its original format, while a lossy format can produce a smaller file size but with a loss of image details. Vector file formats store the image as a series of vectors. Increasing the size of an image does not therefore necessarily increase the size of the image file, and two different images of the same size may, in this case, result in different file sizes. Both Portable Document Format (PDF) and Encapsulated PostScript (EPS) are forms of vector file formats. Vector formatted images are commonly used in publishing, and you may be asked to supply images in this format for a journal. However, they are not suitable for astronomical imaging.

If the detector you are using doesn't allow files to be saved as FITS, then it is likely that it will be saved either as a proprietary format, one of several hundred in use, or as one of the standard formats.

FITS is always the preferred format, being the standard format for astronomy: lossless, uncompressed, and containing a header with information regarding the image. RAW is a similar format to FITS and is used with DSLR cameras. RAW is uncompressed and like FITS has a header. Most DSLRs use a proprietary form of RAW, for example, NEF or CRW, but effectively there is very little difference between the formats, most being based on the TIFF image format with the addition of a header and other additional data structures.

Tagged Image File Format (TIFF) is an uncompressed image format developed in the mid-1980s for use with imaging systems such as scanners. Although not so widely used as in the past, TIFF is still found as a common output option for many imaging software packages.

Like TIFF, BMP (BMP is not an acronym) is an uncompressed image format developed by Microsoft. BMP is a common output file format and is a suitable alternative if TIFF, FITS, or RAW is unavailable. The most common image format in use universally is the Joint Photographic Experts Group (JPEG) format. JPEG is a lossy compressed image format, although some applications do allow for zero compression, which would make a JPEG file lossless. However, as JPEG does not support a significant header, there is no way to determine the degree of loss. JPEG is an extremely common file format, and most image applications can read and write in JPEG. However, it should be considered only if all of the other file formats are unavailable, which in a good astronomical imaging setup should never be the case.

Chapter 8
Imaging

Abstract Astronomical imaging using a telescope and CCD camera is the foundation of modern astronomy. We show through practical examples how to plan, take, and calibrate astronomical images for use in science. We work through the operation of the telescope, including pointing, focusing, and camera operation. We discuss problems associated with nonstellar sources such as galaxies, nebulae, and planets.

8.1 Introduction

Imaging is one of the four cornerstones of observational astronomy. The others are photometry, astrometry, and spectrography. We can consider these practices to answer the following questions: What does it look like? How bright is it? Where is it? What is it made of? Imaging is an important part of photometry, astrometry, and spectrography, as it is the collection of photons that makes these aspects possible. When undertaking imaging, we need to understand what we are trying to achieve. If we wish to perform photometry, then we need to keep the camera within its linear range, which is not so important with astrometry, although we should avoid saturating the target. With spectrography, we will be using a spectrograph rather than an imaging camera, and while we are unlikely to overexpose, we will need to undertake careful guiding to keep the target within the spectrograph's slit. If we are imaging to determine the morphology of a target, then a long integration time may be needed whilst avoiding saturation and taking multiple images, which are combined and manipulated to improve the visualisation of the target.

First light is the term used for the point where a telescope takes its first image, the commissioning image. Although it is unlikely that you will be the first to use the telescope, this might be the very first time you have used a telescope to gather photons from an astronomical target. The previous chapters should have prepared you for your first observations, and this chapter will guide you through that first observation from planning to data reduction. The chapters that follow this one assume that you have successfully completed the practicals that follow in this chapter, or at the very least have experience in using astronomical telescopes and CCDs.

© Springer Nature Switzerland AG 2020

M. Gallaway, *An Introduction to Observational Astrophysics*,
Undergraduate Lecture Notes in Physics,
https://doi.org/10.1007/978-3-030-43551-6_8

103

8.2   Planning

As stated previously, telescope time is a precious commodity, and you should plan your evening's observing in advance to get the maximum benefit out of the telescope.

If you are observing as part of a practical, you should read the practical in advance and understand its aims and processes. Reading or rereading the practical in the dome is a waste of time. You should know what telescope you will be using and its optical characteristics, especially the pointing limitations, the limiting magnitude, and the field of view, as well as the range of filters available.

You will have to choose your target from a supplied list of options or select a target that meets a set of criteria, or perhaps you will have been explicitly assigned a target. You will need to ensure that the target will be visible at the time and date you will be observing and that it is within the limiting magnitude of the telescope. Remember that for extended objects, the quoted magnitude is an integrated magnitude. If you know the approximate area of the object, you can get an idea of its point source magnitude, which you will need to know to determine whether the target is bright enough to be observed as well as to gauge the exposure time. You should write down the target's name, the RA and Dec in J2000, and its magnitude in the band you are observing. If you are not asked to target a specific object, then it is always advisable to have a backup object. If the Moon is going to be up, it is best to avoid the region of the sky it is occupying. Some telescope control applications have a Lorentzian Moon avoidance function built in, which warns you if the telescope is pointing too close to the Moon, the acceptable distance depending on the phase. If the Moon is full and you are doing photometry or deep sky imaging, you may wish to observe at a later date, advice that also applies if there is a bright solar system object nearby.

For each target, you should produce a finder chart, which will help you confirm that the target you are looking at is the target you want to be looking at. Most star fields are indistinguishable from one another. Even if your target is obvious, such as a galaxy, having a finder chart that encompasses a larger area than the telescope's field of view helps. If the telescope moves close to the target but not close enough to put it in the field of view, a finder chart can help in finding where the telescope is pointing and how to move it so that your target is in the centre of the field.

There are several finder chart tools to help you find your target. Perhaps the most popular in the professional community is Aladdin, which can be accessed via SIMBAD. Additionally, DS9 has an image server option that allows the downloading of survey images. The DS9 image servers can be found in Analysis ≫ Image Server, where a number of image servers are listed; we suggest using one of the DSS (Digital Sky Survey) servers, as they offer mostly optical images. You can search by object name or coordinates and can also determine the image dimensions. Before you print out the finder image, I suggest inverting it so that the stars appear black against a white background, as this makes annotation easier. Mark the target on the finder chart and mark the orientation of the axis. An example of a DS9 finder chart is shown in Fig. 8.1.

Fig. 8.1 Example of a simple finder chart using the ESO-DSS image server and DS9. Note the orientation arrows. (Image European Southern Observatory)

The best time to image an object is when it crosses the meridian. At that point, you are seeing the object through the minimum amount of atmosphere for that object for that night, which will reduce the need to refocus. If you try to image each object as it passes through the same part of the sky, it will reduce the amount of time your telescope is slewing. Telescope processes that involve using the telescope when it is not imaging, such as downloading, slewing, swapping filters, and focusing are known as overheads. Always try to minimise the ratio of observing time to overhead time. Although you can work out the meridian crossing times for each object you intend to observe, it is time-consuming and is most easily done with a piece of planetarium software such as Stellarium or The Sky. Open the planetarium program and set the time and date for the start of your observations. Using the software, identify where in the sky your targets are and work out the best order in which to observe them, keeping in mind that some will be setting while others will be rising.

8.3   In-Dome Pre-observation Preparation

Before you make your first observation proper, you will need to take calibration frames, check the pointing of the telescope, and focus up. These are a vital part of the observation process, and their importance should not be overlooked.

8.3.1   Calibration Frames

Before you leave the observatory, you will need to create your calibration files, which were discussed in Sect. 7.5. You may do this either before or after imaging. If you arrive and it is cloudy, you have an ideal opportunity to take calibrations frames, while if it is clear, then take the opportunity go to straight to imaging and take calibration frames at the end. A summary of the calibration process is shown in Fig. 8.2. As discussed earlier, many observatories like to take **sky flats**, which use the twilight sky as their light source. However, for a small observatory, the amount of time to get sky flats for narrowband filters is long, and it might not be possible to take all the required sky flats in one session. An alternative is **dome flats**, for which the inside of the observatory dome, illuminated with the dome lights, is used as the light source. Often, the telescope should point at a part of the dome that has been specially prepared for flat fielding, for example, by the use of a flat white screen. A more modern process is to use the electroluminescent screen designed for the instrument you are using. These are designed to give a virtually constant level of illumination. Although it may not be as flat a source as the twilight sky, it is very stable and yields very flat master flats.

In order to take your calibration frames, you will need to cool the camera down to working temperature and take a series of bias frames, at the same binning as you expect to (or have done) during your observing run. You need multiple biases to make a master bias; eight should be enough. Depending on what software you will be using later, it is advisable to name the files such that they can be correctly identified and are postfixed with a number. For example, Bias$-4 \times 4 - 1$.fit, Bias$-4 \times 4 - 2$, Bias$-4 \times 4 - 3$ would be an acceptable naming sequence, where 4×4 is the binning, and the number is the frame number. Most camera control softwares should do this for you, but you might need to set it up first. You may also need to set the frame type card in the header to "Bias" if the camera hasn't done this already.

You will also need to do the same with your dark frames, although these will also be exposure specific. You should already have some idea from your planning what exposure times you are going to use, but you should have a reasonable range of times, say 1, 5, 10, 15, 30, 60, 120, 240, and 300 s. As with the bias frames, you should take a series of dark frames for each exposure time you will use. If your telescope is small enough to have a lens cap, I suggest putting this on when taking darks, as well as turning the lights off, which will reduce the chance of stray light. If you can set the camera software to take a series of darks automatically, as you can in MaxImDL

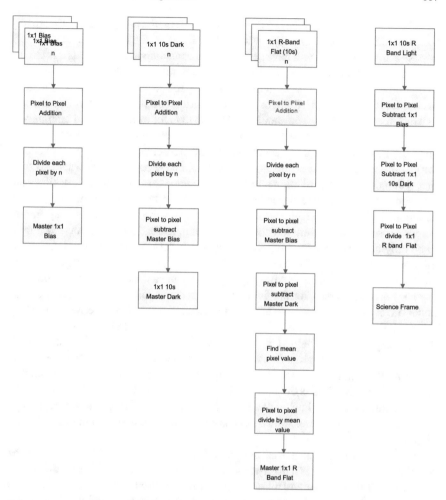

Fig. 8.2 Flow charts showing the processes required to make master bias, dark and flat frames, and the process to convert a light frame to a science frame

in the Autosave tab of the camera control window, it might be a good time to take a break while the camera takes the frames.

In best practice, you should also take multiple flat fields for all the filters you intend to use. However, taking good flats is challenging and time-consuming, so you may wish to use the flats supplied by the observatory staff, who might also be able to supply bias and dark frames if time is short. If, however, you wish to take your own, you will need to take multiple flats for all the filters you are using during your run. If you use supplied flats, ensure that you know whether they are master or single frames.

Whether you are taking flats by dome or by twilight, the two processes are broadly the same. In the case of sky flats, you will be imaging the part of the sky directly opposite the Sun (which will have set), as this is considered the flattest. If it is cloudy or foggy in the area of the sky where you intend to take the flats, then you will need to do dome flats. If you intend to do dome flats, follow the procedures outlined by the observatory staff, which might require pointing at a particular part of the dome or using special lighting. You should take dome flats for every filter and binning you intend to use and, if possible, as many of each as you can in the time you have. A good flat fills each pixel up to about half its full well capacity, about 30,000 counts for a 16-bit camera. Your observatory staff should have some idea of what exposure time you need for each binning and filter to achieve this (note that these times are also camera and telescope dependent). If you do not have this information, it will be a matter of trial and error to find the correct exposure time. Do not keep frames whose exposure times were too short or too long, as that leads to errors. You need to be only at around half the full well capacity. It does not matter if you are a few thousand counts out, as long as the exposure time for each set is consistent when you are taking them. If you are taking sky flats and you can see stars in the frame, the sky has become too dark for flat fielding. If your flat exposure time is short, perhaps under 5 s, it means your source is too bright, and you might have errors caused by the shutter in the flat field. A neutral density filter or neutral density film placed in the optical pathway can help to increase the exposure time without affecting the flat.

Once you have calibration frames, you will need to create **master frames**. These are calibration frames produced by taking the mean pixel value of a number of calibration frames. This is done in order to reduce noise that will likely be present in single frames. Depending on the level of error you are willing to tolerate, you can use single frames as master frames; this is especially true if you do not intend to do photometry.

In order to obtain your master bias calibration frames, you will need to find the mean of the frames; this is a pixel-to-pixel mean and not the mean across the frame. Be very careful when making masters. All the frames used to make the master must be of the same size, i.e., have the same binning (for dark, bias, and flats), the same exposure time (for dark frames), and the same filter (for light frames). If you do not have the same binning level, the process will usually fail, but such is not the case with filters and exposure time. If you make a mistake at this point and it goes unnoticed, it will affect all your images. Most image processing software has some form of image mathematics. If yours does not, then it is reasonably straightforward to create master frames using Python, although you will need the PyFITS and Numpy modules installed.

To turn your bias frames into master bias frames, pixel to pixel, co-add all the bias frames of the binning and then divide every pixel by the number of co-added frames.

For the master dark frames, pixel-to-pixel co-add all the frames with the same binning and exposure time. Divide every pixel in the resultant frame by the number of co-add frames. From the result of this pixel-to-pixel mathematics, remove the bias master that has the corresponding binning. The result is a master dark for that binning and exposure time.

To create master flats pixel to pixel, co-add all the frames with the same binning and filter. Divide every pixel in the resultant frame by the number of co-add frames. From the result of this pixel-to-pixel process, remove the bias master that has the corresponding binning and the master dark with the corresponding exposure and time binning. Find the mean pixel value within the frame and divide every pixel by that value in order to normalise it. The result is a master flat for that filter and binning.

Applying the masters to light frames is straightforward, and your imaging processing software should be able to do this for you. From your light frame, pixel-to-pixel subtract the master bias with the same binning; then from this subtract the master dark with the correct exposure time and binning. Finally, pixel-to-pixel divide the resultant frame by the master flat with the correct binning and filter. The result is the science frame.

8.3.2 Pointing

Once you have all your calibration frames done, open the dome (assuming that it is dark outside). Some telescopes require that you align to two or three bright stars. If you need to do this, do it now. Your observatory staff should be able to help you perform a pointing alignment if you have not done one before. Even if you don't need to use alignment stars, it is always a good idea to check the alignment by slewing to a bright star and taking a short exposure. Centre the star in the image, and if the telescope has this ability, sync the telescope to the star. Syncing recalibrates the encoders to match the position of the target star. Syncing to the wrong target can cause problems, so make sure that the star you are syncing to is the one you think it is. If you cannot see the star, then slew around in small steps, taking images until either you can see the star or it is clear that your telescope is out of alignment. Note that you might be able to see some stray light coming from the star at the edges of the frame before you see the actual star, which can provide a hint as to the star's location relative to the frame.

If you are unable to find the target star within a reasonable distance, use the finder scope. Most telescopes have a small telescope with a wide field of view co-mounted on the main telescope. The finder scope should have cross-hairs, or if you are lucky, an illuminated dot in the centre. Try placing your star in the centre of the finder scope. If you can't see the star in the finder, you have a more serious problem, but most such difficulties are easy to resolve.

If you cannot find the target star, there are several possible reasons. You may have an incorrect RA and Dec. The time, date, or location could be incorrect on the telescope controller. The telescope could have been incorrectly parked or the encoders that tell the controller the position of the telescope in Alt-Az could be offset. It should be easy to check that the time and position are correct, make sure that you are using the time without daylight saving, which is often the cause of errors.

If you can, synchronise the telescope to an RA and Dec; in addition to syncing with a known star, a process known as **plate solving** may help. Plate solving involves automatically identifying the stars in the image and using their relative positions matched against stellar catalogues to determine the centre of the image, which can then be used to synchronise the telescope. A number of telescope control packages, including CCDsoft and MaximDL, have this feature. It is also available as online services from astrometry.net. Note that if the telescope is a long way out of focus, most plate solving programs will struggle, as they will be unable to identify stars in the frame, in which case you may wish to jump to the focusing section of this practical and then return to aligning the telescope.

If the time and location are correct and plate solving fails to correct the pointing problem, try to **home** the telescope if it has this feature. Homing moves the telescope to the absolute zero position of the encoders. If you have a minor problem with the encoders, this may resolve it. If this fails, try parking the telescope. This will put the telescope into a known position. If there is a pointing problem, you might find that the telescope does not return to the park position. If that is the case, unlock the telescope and move it to its park position by hand, although you should check with the observatory staff beforehand whether that is an approved procedure.

If none of these processes clears the pointing problem, then you may wish to escalate the problem to the observatory staff.

8.3.3 Focusing

There is a tendency for users new to telescopes to struggle with focusing. However, once you know the correct technique, focusing is straightforward. Focusing is achieved by moving the focal plane of the telescope so that it is coplanar with the CCD or retina. Focusing may be accomplished by either moving the principal optical component, i.e., the mirror, or moving the detector using a focuser, which is a mechanical or electromechanical device that moves the position of the sensor. If you have the ability to move the mirror and have use of a focuser, use only the focuser. The only reason you should need the mirror focus is for changing the detector, and in most cases, this should be done only by observatory staff. It is likely that you will have to check and adjust the focus several times during the night. This will be partly due to the contraction or expansion of the telescope due to temperature changes, but also to changes in air mass and the slight movement of the camera as the telescope is moved between horizon and zenith. The air mass is the path length that the light takes through the atmosphere, with an object at the zenith having an air mass of 1.0. Air mass can be calculated from the angular distance that an object lies from the zenith, called the zenith distance and denoted by z. The air mass X can be found from the sec of z. Before I take a science image, I always like to take a short focusing image if I have the time, just long enough to check that the field is in focus and the telescope is pointing at the correct location. Most of the time, no refocus is needed, but it is better to waste a few 10 s exposures than several 300 s ones.

Independent of the focuser design, the procedure for focusing is the same. Initially, it would seem that a star, being a point source at infinity, should occupy only one pixel, which is not the case. The light is spread out over several pixels by a combination of diffraction within the optics and atmospheric turbulence. The amount by which the light is spread is known as the **seeing**; if you plot pixel position against count, you will find that plot, for a focused star, appears Gaussian. This plot is the point spread function **PSF** of the star. A star in focus appears as a round, solid, and small object on the CCD, whilst an out of focus star will have an extended, often doughnut-like, structure caused by the obstruction of the secondary.

Many amateurs use a diffraction mask to assist in focusing. This is a grid structure often made of cardboard or plastic that is placed in the aperture of the telescope. The mask makes a diffraction pattern that is distinctive when the telescope is in focus. The most common mask in use is the **Bahtinov mask**, named after its inventor. The mask produces three diffraction spikes around a bright source, normally a star. As the telescope is focused, the three spikes begin to intercept at the same point, which should be the star that is being focused on. For large telescopes, over about 0.4 m, the Bahtinov mask can become troublesome to use, and many telescopes over this size have astronomical cameras rather the digital SLRs and autofocusers. Consequently, Bahtinov masks tend not to be used on larger telescopes (Fig. 8.3).

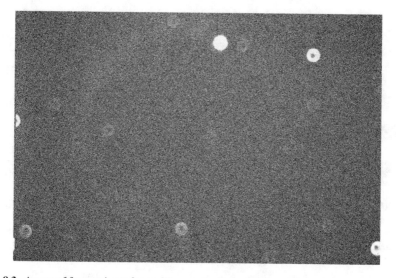

Fig. 8.3 An out of focus science frame. We can clearly see that instead of point sources, we have the characteristic doughnut caused by the obstruction of the pathway by the secondary. The object near the *top centre* is, in fact, bright enough to fill in, but it, too, is out of focus. (Image courtesy University of Hertfordshire)

When you start up, you will need to perform a focus. This is done by pointing the telescope at a bright star and taking an image (look through the telescope if you are using just an eyepiece). Then move the focuser in one direction, either out or in, and take a second image. If the image of the star is getting smaller, you are moving the focuser in the right direction. If it is not getting smaller, move the focuser in the opposite direction. Repeat this cycle until the star is as small as you can get it.

There is a tendency for bright stars to fill in when focusing, so when you have finished focusing on the bright star, move on to a fainter star and repeat the cycle until all the stars in the field are focused. Note that you will need to refocus throughout the night, as the altitude of the target, the air temperature, and possibly the filters used, can all change the focal point.

Your camera control may have some focusing tools that can help your initial focus. For example, it may have the ability to plot the PSF of the star you are focusing on. If you have this feature, it will be easier to determine when the star is in focus, which will occur when the peak is at its minimum diameter. Also, you might have the ability to download only a portion of the CCD, a feature known as sub-framing. This enables you to take faster images of the object you are focusing on, but do not forget to turn sub-framing off when you have finished. Lastly, if you have a powered focuser, you may have the ability to autofocus. The autofocuser moves the focuser in and out until the Full-Width Half Maximum (FWHM) of the star's eps is minimised. The FWHM is a parameter often encountered in physics; it is the width of a peak measured at half the peak's height (Fig. 8.4).

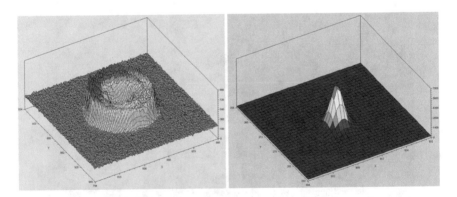

Fig. 8.4 The point spread function of an out of focus star (*left*) and in focus star (*right*). The doughnut PSF structure can be clearly seen, although in the image this star has "filled in" the doughnut. This contrasts to the eps on the *right*, which shows a high, near Gaussian profile, with a distinctive peak indicating that a good focus has been achieved. (Image courtesy University of Hertfordshire)

8.4 Guiding

In most cases, the tracking on the mount you are using will not be good enough to undertake very long exposures without some kind of trailing. Many CCD cameras come with a secondary CCD built in or an accessory that consists of a small, low-cost CCD and a small mirror or prism to guide and focus a portion of the image onto the CCD. These are known as off-axis guiders. Alternatively, your telescope may use a separate coaxial telescope and CCD for guiding. With your camera control software, you should be able to access the guider CCD. Either this will automatically pick a star on which to guide, or you will have to pick a star. If your field is very crowded or contains a large galaxy or nebula, the autoguider may have problems identifying a suitable candidate star, in which case you may have to select one for it. The functionality of your guider will depend on the software you are using, so once again, check with the observatory staff before using this feature. In some cases, you will need the autoguider. For very short exposures, the amount a star moves may be less than one pixel. In this case, autoguiding may be unnecessary, whilst for longer exposures, the need for autoguiding will depend on the tracking quality of the mount. Since telescope mounts are designed for sidereal tracking, when you are imaging solar system objects, which do not move sidereally, you will be able to take only short exposures before you see the object's trail, even with autoguiding.

If an autoguider is unavailable or you do not wish to use it for some reason or it is struggling to identify a suitable guide object, an alternative to tracking is stacking. We will be discussing stacking in more detail later. However, broadly speaking, it is a technique whereby a single exposure is generated by the summation of several shorter exposures.

For each of your targets, you will have either to set up the autoguider or, if it automatically picks a guide star, to check that it is indeed guiding. Generally, changing a filter should not be a problem for the guider. Most telescope control software reports guider data as it exposes, and you may wish to abort the exposure if guiding is lost for than a few tens of seconds.

If you can adjust the exposure time of the guide CCD, it is often useful to increase it in order either to make the guide stars more prominent or to increase the rate of correction. Beware of setting the exposure time too short, as that can cause problems with the control computer.

8.5 Target-Specific Problems

In this section, we will discuss some of the challenges specific to certain target types. Remember that there is a fundamental difference between science images and the images you see on the covers of books and glossy science magazines. The processing used to make those images pleasing to the eye often means that the pixel values are no longer linear. Aside from applying calibration frames, you should not perform

Fig. 8.5 A clear image of the open cluster M26. The *left-hand image* shows how it would normally be displayed, whilst the *right-hand image* is stretched. The pixel range displayed is reduced, so that faint stars are now displayed without making the brighter stars too bright. Stretching in this way affects only how the image is displayed and does not change the pixel values. (Image courtesy University of Hertfordshire)

any action that is likely to change the linearity of the frame, for example by applying a Gaussian filter, unless you intend not to make any measurements on the image. You may wish to scale or **stretch** the image, which will make the fainter structure more visible. Stretching an image changes the way it is displayed. A modern CCD has a range of pixel values far in excess of what can be detected by the human eye when viewed as a greyscale image, a problem that also applies to most monitors. By stretching you display only the pixel range that is of interest, which makes the detail more visible. Figure 8.5 shows unstretched and stretched versions of the same image of M26. It is clear that we can see many more stars after the image has been stretched. This changes only how the image is displayed and does not alter the pixel values, so the frame remains scientifically useful.

Most applications perform stretching using a frequency plot showing the distribution of the pixel values. There will be a large number of pixels with a low count, representing the background, with few pixels at higher levels. With most applications, the range of pixel values displayed is set by either a pair of drag bars on the plot or a pair of sliders. By sliding the bars, you can enhance the display of fainter details. Typically, this is done by moving the minimum value near the centre of the background peak and the maximum bar to a lower value, with the best location often being slightly right of the background peak. Many camera control applications remember the stretch settings, but the best settings depend on a number of factors. If you take an image in one band and perform a stretch and then change the filter, do not forget to change the stretch setting. Otherwise, you may not see the target in the image.

8.5.1 Imaging Open Clusters

Open clusters, also known as galactic clusters, are sites of the most recent star formation in the galaxy, and as the spiral arms are delineated by star formation, they are always found within the confines of the Milky Way. Hence, we do not find open clusters in, for example, Ursa Major, but we find many in Cassiopeia. Open clusters form from collapsing molecular clouds that have become dense enough and cold enough that gravity overcomes the thermal motion, turbulence, and magnetic fields that can prevent collapse. However, do not be fooled by science fiction shows: the density of these clouds is still lower than any vacuum ever achieved on Earth!

As the cloud collapses, local, smaller regions become superdense, and these further collapse into stars. Over time, radiation pressure, stellar winds, and supernovae from the more massive stars formed within the cluster strip the cluster of the gas and dust that was not incorporated into the stars. It is this dust and gas that makes up the bulk of the mass of the cluster, not, as you might expect, the stars. In fact, star formation is quite inefficient, with only about 1–3% of the mass of the initial cloud becoming stars. With the stripping of this gas, the cluster becomes gravitationally unbound and begins to break up, with the stars becoming members of the galactic disk. Depending on the mass of the cluster, this can take anywhere from tens of millions to a billion years.

As a point of interest, open clusters appear to be mass segregated, with more massive stars found near the centre of the cluster and lower-mass stars near the edge. Why this happens is a matter of debate. Models suggest that over time, gravitational interaction will segregate the cluster. However, large-scale star formation models suggest that the most massive stars tend to form in the densest part of the cloud, which is typically at the centre. Perhaps it is more likely that both processes contribute to the overall distribution of stars within the cluster. Nonetheless, open clusters are some of the less challenging objects for the proto-astrophysicist to image.

The principal problem undergraduates encounter when exposing an open cluster is overexposing and therefore saturating the brighter stars, especially when an exposure time is not adjusted between bands. Overexposure, or in fact entering the nonlinear region of the CCD, that occurs before saturation can render the image effectively useless for science, which is not immediately obvious when an image is displayed. It is important to understand a number of effects. Firstly, the size of the star on the image is not representative of its brightness, as the processes that cause the light to be spread over multiple pixels, diffraction and atmospheric seeing, affect all stars in the field identically, and the star's apparent size is therefore independent of the brightness of the source. Visual detection of saturation is problematic, as computer monitors are unable to display the full pixel range in grey levels, and even if they could, the human eye would be unable to distinguish between the grey levels over the entire range. To resolve this problem, you will have to perform an image stretch (also known as scaling). This has nothing to do with changing the shape or size of the image; rather, it is how the image is displayed. In general, stretching does little to

Fig. 8.6 Inverted B-band image of the open cluster M26. (Image courtesy University of Hertfordshire)

improve the detection of saturated pixels, but it does have an important application
that will be discussed later.

How is saturation detected? The simplest method is to do a surface plot or a
slice of some of the brighter stars in the field. Most astronomical imaging or image
processing applications should be able to perform this function, including MaxIm
DL, APT, and SalsaJ. To illustrate, Fig. 8.6 shows a B-band image of the open cluster
M26. The image is inverted, i.e., black and white are reversed, in order to make it
clearer. From visual inspection, none of the stars appear saturated. However, when
we perform a surface plot of a bright star within the field, Fig. 8.7, we see that it
does not have the characteristic Gaussian shape we would expect; rather, it is domed.
This indicates that the exposure was long enough for the pixel's response to become
nonlinear. Note that if the top of the peak were flat, that would indicate saturation,
although of course, the pixels would have gone nonlinear before that happened. The
smaller peak in Fig. 8.7 is another, dimmer, star, caught within the plot region. As
can be seen, this is broadly Gaussian and therefore acceptable. Figure 8.8 shows the
same region as Fig. 8.7, but in this case, it is a slice instead of a region plot. However,
we can see the nonlinear nature of the bright star within the plot and the Gaussian
nature of the dimmer star also intersected by the slice.

Ideally, you want the brightest star in the image to be just inside the linear range
of the CCD. If you have a saturated star in the field, you may wish to retake the
light frame with reduced exposure time. If you know the count and the magnitude
of an unsaturated star (see the chapter on photometry for how to measure the count)
as well as the magnitude of the saturated star, you can determine the count of the
saturated star if it was unsaturated using (8.1), where M_{ref} and M_i are the magnitude

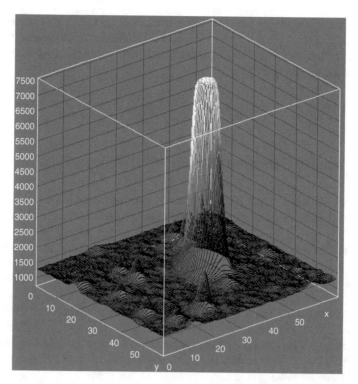

Fig. 8.7 Surface plot of a bright star within Fig. 8.6 using SalsaJ. The dome-like structure of the plot suggests that the star lies within the nonlinear region of the CCD. A flat top, as opposed to a dome, would indicate saturation. Note the smaller peak, which is much more Gaussian-like than the larger peak. This is a dimmer star within the region of the plot, and its Gaussian nature indicates that its pixel values lie completely within the linear region of the CCD

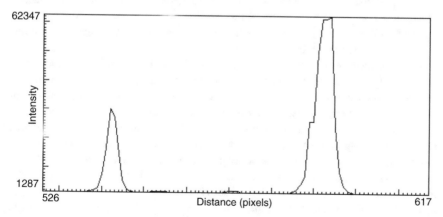

Fig. 8.8 A slice though a bright star within Fig. 8.6 using SalsaJ. The large peak is clearly non-Gaussian, indicating that it lies within the nonlinear region of the CCD (a flattened peak would also indicate saturation). The small peak is from another star also in the slice. This appears broadly Gaussian in nature as we would expect and hence lies in the linear range of the CCD

of the unsaturated and saturated star respectively, and I_{ref} and I_i are their integrated counts; you can scale your exposure time such that in the revised exposure, the star is unsaturated. Note that this is band specific and to some extent spectral type specific, and uncertainties can be high:

$$I_{ref} \times 10^{\frac{M_{ref}-M_i}{2.5}} = I_i \tag{8.1}$$

8.5.2 Imaging Globular Clusters

Globular clusters are very different objects from open clusters. They are old, very dense, found exclusively in the galactic halo, and in general, are considerably more massive than open clusters. Unlike open clusters, no star formation has ever been detected within a globular cluster, and the stars contained within have lower levels of elements with atomic numbers greater than two (which in astronomy parlance are known as metals, with the ratio between hydrogen and a metals being its **metallicity**), indicating that they formed early in the evolution of the galaxy. These low metallicity stars are known as **population II** stars, while the younger, higher metallicity, stars such as those found in open clusters and also including stars such as the Sun, belong to **population I**. There is a possibility that ancient, ultralow metallicity **population III** stars exist, most likely as very low mass, and therefore slow-burning, M-Dwarfs, but these have yet to be detected.

The challenges for imaging a globular cluster are very different from those of an open cluster. As globular clusters are in the halo, they lie at much greater distances than many open clusters, and any very bright star is therefore likely to be a foreground star rather than a cluster member. However, the globular cluster is a very dense ball of stars, perhaps up to several hundreds of thousands of stars with an average density of perhaps one per cubic parsec. Although this does not sound significant, near the centre of the cluster, the average distance between stars may decrease to a few hundred astronomical units.

Consider the line of sight as we pass from the edge of the cluster towards the centre. Near the edge, the number of stars within the line of sight is low, and individual stars are seen. However, as we move towards the centre, the number of stars in the line of sight increases to the point where we are no longer able to resolve individual stars.

The secret to imaging a globular cluster, therefore, is to enhance the detectable stars near the edge whilst not saturating the core, where we are unable to resolve individual stars. This we can achieve by taking multiple images and stacking them in a nonlinear fashion. This can be used for general imaging and astrometry but not for photometry, as it makes the pixel values nonlinear.

Figure 8.9 shows a science frame of the globular cluster M3 taken in clear (i.e., without a filter). The exposure time was 60 s, and no other action than the removal of the bias, dark, and flat frames has been undertaken except for inverting the image. We can clearly see individual stars near the edge of the cluster, but the centre *appears* saturated.

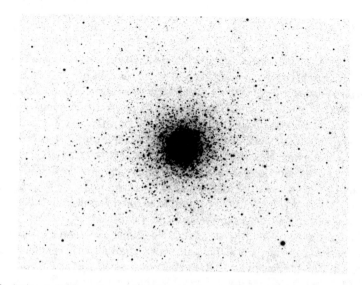

Fig. 8.9 An inverted science frame of the globular cluster M3 taken in clear. This image has yet to undergo any post-production correction to improve the visibility of the stars near the *centre*. (Image courtesy University of Hertfordshire)

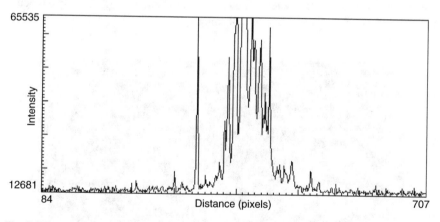

Fig. 8.10 A slice though Fig. 8.9 using SalsaJ. We can see that near the centre of the globular cluster there are many saturated stars The sharp peak to the left of the main body is caused by a cosmic ray

Figure 8.10 shows a slice through Fig. 8.9. The apparent rise in the background near to the centre of the field is in fact due to the stars in the centre of the cluster merging in the image. We can also see that there are a number of regions in the centre that are saturated. The sharp peak located to the left is due to a cosmic ray.

8.5.3 Imaging Galaxies

In many ways, the challenges of imaging globular clusters apply to galaxies. Spiral galaxies have bright cores comprising a dense congregation of old stars surrounded by a thin disk of stars and gas with the spiral structure being delineated by the regions where stars are forming. Elliptical Galaxies are no longer star-forming (or are doing so at a very low rate) and have become virialised. In essence, this means that the kinetic energy of the stars within the galaxy has become evenly distributed. There is no disk, as with a spiral galaxy, just a dense ellipse of stars. Imaging elliptical galaxies is just a matter of trying to keep the galaxy in the linear region of the CCD. Remember that the most quoted magnitudes for galaxies are the integrated magnitude and not the surface brightness, it is the former you need to use to determine the exposure time. Spiral galaxies can have a considerable structure which is often very faint. This leads to a dichotomy, as the nucleus is very bright necessitating a short exposure, whilst the arms and dust lanes are faint and require a long exposure. In general, it is better to stay within the linear range of the CCD but take multiple images and mathematically manipulate them to bring out the structure. As spiral galaxies are star-forming they can produce a considerable amount of $H\alpha$ emission which is limited to the location of the star formation, i.e. in the arms. Hence, the arms tend to be brighter in $H\alpha$ than the nucleus. The inclusion, thereto, of an $H\alpha$ filter when imaging can help delineate the arms. Figure 8.11 shows the effect of imaging a star-forming galaxy in $H\alpha$.

Fig. 8.11 An image of the spiral galaxy M81. This was taken in three filters with *blue* being a Johnson B filter, *green* a Johnson V filter and *red* an $H\alpha$ filter. The use of an $H\alpha$ filter enables the spiral arms, the regions of star formation, to be distinguished. (Image courtesy University of Hertfordshire)

8.5.4 Planetary, Emission, and Dark Nebulae

Planetary nebulae (PNe), named for their planet-like appearance in early telescopes, are the shells of stars that have been discarded during the red giant phase. The hot progenitor star, often a white dwarf, ionizes the expanding envelope, which leads to the emission of forbidden lines. Planetary nebulae have complex structures that are very poorly understood. Many exhibit an hourglass-like structure and often appear circular or elliptical in small telescopes. Planetary nebulae are rare in the Milky Way despite a large number of known possible progenitor stars, suggesting that they are very short lived, perhaps only about 10,000 years.

Planetary nebulae can be challenging if their angular size is small, as many are, since that makes resolving detail difficult. Although it is possible to image PNe in broadband filters, their very nature lends them to narrowband imaging, as that is where the bulk of the emission lies. The use of narrowband filters will also reduce the sky contribution, improving the signal-to-noise ratio. However, for very faint PNe, very long exposures are needed, which can be achieved by taking multiple images and stacking them later. If they are very small, the introduction of a Barlow or similar device into the optical pathway can increase the plate scale.

Emission nebulae, of which PNe are a subclass, are clouds of gas often star-forming and with embedded stars. These stars ionise the gas, which then deionises, emitting light. Just as in the case of PNe, emission nebulae can be very diffuse and faint but can also be very bright in some regions, especially if there are massive stars illuminating the cloud. One such example is M42, also known as the Orion nebula, as illustrated by Fig. 8.12. M42 is also a **diffuse nebula**, having no well-defined boundary, as can be seen. In astrophysics, diatomic hydrogen is known as hydrogen and denoted by H, disassociated hydrogen is known as **HI**, whilst ionised hydrogen is known as **HII**. Hence M42, which is mostly ionised, is an HII region, as well as a diffuse and an emission nebula.

The core of M42 is illuminated by a number of very massive stars, which can cause the saturation of the core. Only by stretching the image, reducing the dominance of the core emission, do we see the fine structure. There are a number of computational filters that can enhance detail, for example a logarithmic stretch, although this will make any results nonphotometric. However, in many cases it is the morphology that interests us, in which case sacrificing the linearity of the image is acceptable.

Dark nebulae are dense cold molecular clouds. Because of their nature, they are visible only in the optical region when located in front of an emission nebula. Although they often appear dark in front of the emission nebula. Dark Nebula can be seen in Fig. 8.12, although perhaps the most famous example is IC434. Hence when you are imaging dark nebulae, the integration time will depend on the brightness of the background source.

Fig. 8.12 An image of the emission nebula M42. This was taken in three narrow-band filters with *blue* an [OIII] filter, *green* an $H\alpha$ filter, and *red* an [SII] filter. The use of narrow-band filters enables subtle features within the nebula to be distinguished. Note the very bright core, which is host to a number of massive stars. (Image courtesy University of Hertfordshire)

8.6 Stacking and Dithering

For unresolved astronomical objects, it is likely you will be producing one image in each band unless you are undertaking time-series photometry, which is discussed in Chap. 10. For resolved objects such as nebulae and galaxies, it is likely that you will require a long integration time to bring out the detail within those objects and to improve the signal-to-noise ratio. Very long integration times pose a number of problems. Firstly, if there is a bright region in the image, for example a foreground star, it may saturate and bloom before you have enough photons from your target. Secondly, the number of cosmic ray strikes on the CCD is a function of time, so as you increase the exposure time, the number of cosmic ray strikes goes up proportionally. Also, the longer you expose, the more likely that something will go wrong, for example a plane or a satellite passing through the shot or tracking errors. Lastly, there just might not be enough good weather or even dark enough skies in one session to complete the full exposure.

 To overcome these problems, astronomers can take multiple frames of the same target and stack them pixel to pixel. This technique is discussed more fully in Chap. 9. A stacked frame is not limited to the same bit constraints as a camera; so, for example, images from a 16-bit camera can be stacked to form a 32-Bit FITS, solving the problem of very bright objects in the field.

Images need to be aligned before they can be stacked unless your telescope has perfect tracking. However, the alignment process enables us to deal with a number of issues. Variations within the CCD (for example bad pixels) can be evened out over the chip by **nodding** or **dithering** the telescope. Dithering is performed by moving the telescope a few arc seconds between exposures. Faults move with the telescope, and so they appear in a different location in each frame. This can be used to minimise their effects.

So we see that when we make our observations, we must be aware of the image postprocessing that will be needed and take a number of exposures at an exposure time appropriate to what is needed, both for the initial image and for later processing.

8.7 Practical 6: The Messier Marathon

The practicals in this chapter are designed to give you the skills to take science grade astronomical images in an efficient manner so that you can utilise your precious clear sky time to the maximum. You will be imaging a wide range of targets that require the implementation of different imaging approaches. You will gain experience not only on the telescope, but also with a range of astronomical software packages.

A common challenge amongst amateur astronomers is the Messier marathon—seeing how many Messier objects one can image in a single night. A successful Messier marathon requires careful planning and preparation, exactly the skills a good observational astronomer needs. If you are a southern hemisphere observer, the Messier marathon can be quite challenging, due to the limited number of southern objects and the low altitude of many of those that are visible. If this is the case, we suggest you use the Caldwell catalogue instead or as an addition. As noted in Chap. 6, the Caldwell catalogue contains a further 109 bright objects that are not in the Messier catalogue.

8.7.1 Aims

In this practical you will attempt to image as many Messier objects as possible in a single observing session. Not only that, you will make colour composite images, meaning that you will need to image in at least three filters. This practical will take careful planning and teamwork if you are to execute it successfully. You will be familiarising yourself with the use of the telescope and the camera to image a wide range of object types, from bright clusters to faint galaxies.

8.7.2 Planning

As mentioned above, to successfully complete this practical, you will need some careful planning. Begin by starting up your planetarium software and setting it to the time and date on which you expect to observe. Bring up the constellation boundaries and names, have a look at what will be visible, and determine whether each object is rising or setting. Check the position of the Moon, if there is one, as for some of your targets, you may need to avoid it.

You will now need to look at the Messier catalogue. You need to observe at least one open cluster, one barred spiral galaxy, an elliptical galaxy, a spiral galaxy, an HII region or supernova remnant, a planetary nebula, and a globular cluster. Using your understanding of what is currently visible, pick the best-placed target of each kind. You will also need backup targets for each class of object.

Once you have selected your targets, you will need to determine the order of observation and, based on their magnitudes, their lengths of exposure. You should also be aware of the angular size of the objects. Some may be too small for your camera to image, while some (for example, M31) will be too large. You should image them in at least three bands. If you have narrow-band filters (such as OIII, Hα, and SII), the HII region and the planetary nebula will benefit from their use. Although the use of narrowband filters will increase integration time, low surface brightness objects that emit strongly in these lines benefit greatly, as the sky tends not to be so bright around forbidden lines (although emission lines are still present), improving your signal-to-noise ratio for these sources.

8.7.3 Getting Ready

Before you leave for the observatory, make sure you have all your documentation, your lab book, something to put your images on, such as a memory stick, and appropriate clothing. It is best to assume that it will be colder than forecast, as you can always take a sweater off, but you can't put on one you don't have!

1. Once you arrive at the observatory, you need to start up the telescope and the instrument. A member of observatory staff (or your tutor) should be able to help you if you don't know how to do this. The first thing you will need to do is start the camera coolers if they are not already on, as it takes a while to bring the camera down to operational temperature.
2. Once the camera is down to the required temperature, you will need to take calibration frames. If you have been recently observing with this telescope and camera and have bias, dark, and flats for the binning, and the exposure times and filters you intend to use, then you can most likely use them instead of taking new ones, although they may limit your exposure times. If your flats are more than a week old, it is normally worth taking new ones. Dark and bias frames are

normally good for a month, as long as there have not been any major changes in the setup of the telescope or new equipment installed in the dome.

3. Once you have taken your calibration frames, if you need them, open the telescope dome and go through the pointing and focus routine. Once this is done, you can move on to imaging.

8.7.4 Imaging

1. For each of your targets you will need to slew to the target using the coordinates or, if you have a GoTo facility, the target's name.
2. Take a single, fairly short, image in clear or the nearest equivalent filter, for example, a Johnson V. Use this to check your focusing.
3. Confirm that the target is in the centre of the field. If it has low surface brightness, the object might not be visible without stretching, and even with stretching it might be challenging to distinguish. If you cannot identify the target, use the stars and your finder target to attempt to locate it.
4. Once you are happy with both the centring of your target and the focusing, turn on and calibrate your guider and check that you are tracking. Remember that guiding is a process requiring communication between camera and mount, while tracking is just the sidereal movement of the telescope.
5. Pick your first filter and start imaging your target. Once the first image is displayed, check that it is not under- or overexposed. You may need to stretch the image. If your exposure is too short, you can work out how much longer you need by examining the pixel value of the brightest target. Pixel value should effectively scale with exposure time, although of course, the bias and pedestal do not. However, the contribution to your pixel value of these components should be quite small, and their nonlinear nature gives you a degree of headroom in scaling up exposure times. For example, if you are exposing for 10 s and your target's maximum pixel count is 10,000, assuming that your maximum linear pixel count is 50,000, you can safely expose for 50 s, because the bias and pedestal, which likely make around 100 counts, remain constant with time; this being the case, the maximum pixel count is more likely to be 49,600 for a 50 s exposure.
6. Once you have done this process for all three filters and you have saved your images and noted your exposure times and file names in the log book, you can move on to the next target. However, you may wish to take multiple images in each band and stack them in software later, as this reduces noise, as we will see in Chaps. 9 and 11. If this is the case, once you have your first set of RGB images (or whatever filter series you are using) and you have confirmed your exposure time, you may wish to use the automation service available in your camera control package. This will take a series of images in multiple filters without requiring your interaction. However, I strongly recommend you always take the first set manually to confirm that the chosen setup is correct. Likewise, you should be

aware of how long it will take to run and be at the telescope before it finishes. Otherwise, the instrument might stand idle when it could be being used.

7. Although not entirely necessary, many amateur astronomers like to take a clear frame in addition to the three coloured filters. The clear is used to form a **luminance frame** in the modified version of the standard RGB colour images, known as an LRGB. LRGB images tend to have a better definition of fine structure, but not all imaging software can produce LRGB images, although MaxImDL can.

8.7.5   Shutdown

Once you have all your light frames, do not forget to take your calibration frames if you haven't already. Make sure you have everything in your lab book that you need, including the details of the telescope and the time you finished imaging.

It is likely that your instrument has a specific shutdown procedure, which you should know. If you do not, please check with a member of the observatory staff. In general, this procedure requires parking the telescope, which puts the instrument in a known position so that when it is restarted, it knows where it is pointing. The dome will need closing, and if it is a tracking dome, it may also need parking. You should shut down the camera and its cooler and turn on any ancillary equipment that you might have turned off, and vice versa. For example, some observatories use dehumidifiers to protect the optics.

8.7.6   Post-imaging

After taking your images, you will need to produce your science frames. In order to do this, you will need to subtract the correct dark frame from each of your flats and then normalise the flat.

1. Subtract the dark frames from your light frames.
2. Divide the result by the flat to produce a science frame.
   Note that since the dark frame includes a bias, it is not strictly necessary to have bias frames. However, if you do not have a dark frame with the correct exposure time for some reason and you need to make one by scaling another dark frame, you will need to subtract the bias first, as it doesn't scale, and then add it back after the scaling. As bias frames are so quick and easy to take and can be useful, I suggest you always take them even if in the end you do not use them. The process of making science frames is discussed in more detail in Sect. 7 in Chap. 9.
3. If you have multiple images of each band, you should stack these to improve the signal-to-noise ratio. However, be aware that when you stack frames, the nature of the stack can destroy the scientific usefulness of the data. For the Messier marathon, this is not an issue, as you are attempting just to produce an image, but

for photometry and spectrography you should stack only via addition, although you may make a mean (but not a median) stack if all your exposures are of the same length. Details of how to align and stack images are presented in Chap. 9.

4. Once you have your three separate frames, four if you intend to do an LRGB, you will need to tidy them up and stretch them before combining them into your colour image. Hot pixels and cosmic rays will appear as unusually coloured pixels when you create the colour image and are easier to deal with at this point using a cloning tool. If your astronomical imaging software has a hot pixel removal tool, it is worth applying it at this point, although I find that often they fail to get them all, especially cosmic rays.

5. Combine the images using your favourite software into an RGB or LRGB image and then colour balance them as discussed in Chap. 9. Once you are happy with your image, export it in TIFF or JPEG format. You may then use standard image manipulation tools such as Photoshop or The Gimp to remove artifacts and crop the image.

8.7.7 Analysis

When you have finished, you should write up your work and include your images, finder charts, and exposure times. During the writeup, pay particular attention to any problems you encountered. Describe what you did to the images, why you did what you did, and how it influenced them. In particular, look at how the signal-to-noise ratio improves when you add calibration frames and science frame stacking.

8.8 Planetary Imaging

It is often thought by students that the easiest objects to image are the planets, because they are so bright, but that is not true for most teaching observatories. In fact, planetary imaging is extremely challenging for the very reason that planets are so bright. Most astronomical cameras cannot cope with the number of photons being received from the brighter planets—Jupiter, Mars, Venus, and Saturn—and they overexpose even with their shortest exposure setting and with a photometric filter in the pathway. Another problem is that the telescope is designed to track not solar system objects but stars. Autoguiding is not of as much help here, as it would need to guide on the planet, which, not being a point source, causes problems with guiding software. Hence, even if we wished to take long exposures, they would suffer from trailing.

An obvious solution to this problem is to reduce the amount of light entering the camera. This could be done either by **stopping down** the telescope, i.e., putting a cover on the front to reduce the aperture size or putting either a twin polarising filter or neutral density filter in the optical pathway. Another challenge is that to see detail on the planet, high magnifications are needed. At high magnifications, seeing becomes a major limiting factor. As we have seen, the turbulent motion of the atmosphere limits the resolution of the telescope, which becomes very obvious with solar system objects. Again we could reduce our exposure time to below the turbulent time scale, which can be smaller than a hundredth of a second, far shorter than most astronomical CCDs can image.

8.8.1 Lucky Imaging

There is, however, a simple solution to these problems that was first implemented by amateur astronomers but is now being picked up by professionals; the use of webcams. Although insensitive and noisy compared to astronomical cameras, their speed and comparatively low cost led to their modification by amateurs into planetary imaging cameras. However, in recent years, low noise astronomical webcams, both colour and black and white, have started to appear on the market, and these are a cost-effective solution to the problems with planetary imaging using a conventional astronomical camera. Professional ultra-high-speed cameras, which may be cooled and have noise and sensitivity levels close to the standard astronomical cameras as well as frame rates as high as a thousand frames a second, are also becoming available, but at considerable cost.

The advantage of using a high-speed, but relatively insensitive, camera is that the frame rate is typically faster than the turbulence time scale. Imaging software, such as Registax, is capable of processing the video file that comes off these cameras frame by frame. By picking only the best-aligned frames, where turbulence is not present, and stacking those frames, ultra-high-resolution imagery is possible, well below the typical seeing and often limited by the diffraction limit of the telescope or the pixel size of the camera. Resolutions of 0.2 arcsec using this technique, known as **lucky imaging**, are common. This is typical of the resolutions achieved at very high altitude observatories such as those in Hawaii and Chile, or with adaptive optics.

There is yet another use for **lucky imagers**, as these devices have become known. Their very high frame rate allows the observation and precise timing of very transient astronomical phenomena that are too short for normal CCD-based astronomical cameras to capture. Examples of these are asteroid occultations, which may be a few seconds long, and lunar meteoroid impacts on the Moon.

Given the very light nature of most lucky imagers, the time taken in removing and then replacing an astronomical camera, the fact that a guide camera is not needed for planetary imaging, and that most guide cameras are mounted on refractors, which are ideal for planetary imaging, there is an advantage in using the guide telescope for planetary imaging. If you have a choice, and you might not, we would suggest going down this route for most planetary imaging.

8.9 Practical 7: The Speed of Light

In 1676, the Danish astronomer Ole Rømer noticed that the eclipses of Io, in front and behind Jupiter, lagged behind what was predicted by orbital mechanics and that the lag seemed to be dependent on the distance of the Earth from the planet. Rømer realised that this implied that the speed of light was finite, and he managed to measure it at close to current estimates. With modern high-speed cameras and increasing computing power, it is possible for you to follow in Ole Rømer's footsteps.

8.9.1 Aims

In this practical, you will image a solar system object using a high-speed astronomical camera or webcam with the goal of identifying the moment in time at which an occultation occurs. By looking at the difference between the predicted moment that it should occur (without accounting for the speed of light) and the actual moment of occurrence, you will then be able to calculate the speed of light.

8.9.2 Planning

This project needs careful planning, as you will identify an occultation or transit that will take place during your observing time. If Jupiter is visible, then the transit of Io is the best candidate, as such transits are common and obvious. If not, you could use a planetary occultation of a star. Try to avoid using the Moon, since although lunar occultations of bright stars are common (planets less so), the Moon is very close and will introduce large errors into your results.

You will need to use a piece of planetarium software to find the uncorrected time of the occultation or transit. Stellarium is excellent for this. Just remember to turn off the option to simulate light speed found in the view window. You will also need to find the distance between the Earth and the solar system object at the point of the occultation or transit. Again Stellarium can do this for you.

You are likely to generate a large video file, so make sure you have some way of getting this off the camera control computer. You will also need to carefully measure the moment the transit is observed, so ensure that you have an excellent well-calibrated timepiece.

8.9.3 Observing

1. It is best to turn up at the observatory well before the transit is due so that you can make sure that everything is set up and that you are in focus and pointing correctly.
2. Start imaging your target about 15 min before the event is due, but there is no need to record at this point. Remember that you will be unable to autoguide the telescope, and even with tracking on, the target will move in the field. Hence, you will need to track the target manually.
3. You should see the two bodies involved in the occultation or transit in the frame and be able to make a reasonable judgment as to how close you are to the event occurring. Start recording when you think you are about 10 min from the transit. Make a careful note of when you started recording. Try to determine accuracy to about a second and the frame rate at which the camera is running. Depending on your object, the event could be delayed by as much as an hour from that reported by Stellarium.
4. Once you are happy that you have captured the event you wished to observe, stop and save the video file. Back it up and transfer it to another medium for removal from the camera control computer before shutting the observatory down.

8.9.4 Analysis

1. Using the video file, play the video until the moment that the transit or occultation occurs. You might need to rewind and play the video a number of times in order to identify the frame number.
2. Dividing the frame number by the frame rate gives the number of seconds after recording started that the event occurred. using your starting time and this figure, along with the uncorrected time and distance from Stellarium, you can calculate the speed of light.
3. As with previous practicals, you should write up your report in standard laboratory style. You should include notes about the camera and software used in addition to any problems you may have encountered. You should also include details of your preparation, especially details concerning the occultation or eclipse and its expected, ucorrected, predicted time. Pay special attention to how your result compares to other more modern methods of determining c, what your uncertainties are, where they come from, and what you would do differently if you were to repeat this experiment.

8.9.5 Remote Observing

As discussed in Chap. 1, remote and robotic observing is becoming the standard for many observatories due to the introduction of low-cost, high-quality GoTo mounts and the standardisation of software driver formats via the ASCOM standard. This allows observatories to mix dome controllers, mounts, cameras, and powered focusers from different suppliers to meet their own needs. This has also led to a wide array of software interfaces for robotic and remote telescope systems. Often in robotic observatories, behind many robotic telescope interfaces lies a programming language known as Robotic (or Remote) Telescope Mark-up Language (RTML). RTML is very similar to HTML and is the actual code that the telescope control computer uses to tell it what to do.

Hence, no two robotic observatories will be operated identically, and the discussion of an individual system falls outside the scope of this book. Among the main reasons that many observatories have moved towards robotic and remote operation are the cost of labour required to supply nighttime support as well as daytime maintenance and the reduction in travel costs to remote sites. To some extent, the benefits of robotic and remote observing are offset by the loss of telescope time when a fault develops, which is more common in these systems due to their complexity. In many situations, these would be solved by an on-site astronomer, and simple reboots of computer systems and cameras are possible to perform remotely, but hard resets (the too common turn-it-on-and-off-again fix beloved by IT professionals) are not. A minor problem for an on-site observer can often become a major one for an off-site observer.

Service mode observing involves another astronomer or telescope operator (TO) using the telescope on your behalf. Service mode is becoming increasing popular amongst research-grade observatories such as Gemini and the VLT, where technical staff are always on hand to take the observations, and flying astronomers to the observatory is becoming increasingly cost-ineffective (in addition to being bad for the environment). As an undergraduate, you are unlikely to encounter service mode operation. However, both remote operation and robotic operation are becoming increasingly commonplace.

As stated earlier in this book, remote observing means that a user has direct control of a telescope, its camera, the dome, and any ancillary equipment such as the filter wheel. This is often undertaken using either a secure remote desktop application built into the operating system or an off-the-shelf package such as VNC. Robotic or queue observing does not invoke direct access to the telescope as in remote observing. In robotic obsesrving, instructions are fed to a computer system that directs the operation of the telescope and runs the client's observations, normally based on a priory algorithm that may or may not have some human input.

For robotic observation, you will need to prepare a detailed plan for your observations and be able to translate that plan into a format that the telescope will understand. This is typically done using user interfaces like the one used by the Liverpool Telescope or iTelescope. As each telescope operator has different interfaces, and even those with common interfaces will have some configuration differences, you must read and understand all the documentation that the operator asks you to read. It will increase your chances of an observation being accepted and successfully run. Many remote telescope operators will ask within the submission process for supporting documents and a written description of what you are trying to achieve and why. Stating that you just want to take pretty pictures is unlikely to get you telescope time.

How you build your plan and the requirements you state for observation will have an impact on when, if ever, your observations are run. Your plan needs to be realistic. If the telescope you are using is being used for research, postgraduate projects, or final year projects, it is likely you will have a low priory to start with. You should write the observing plan with the following in mind; what is the minimum I need to get my observations done and how can I improve my chances of it happening? Unless you are doing very high photometric resolution photometry, you are not going to need a completely moonless night or photometric weather; requiring such conditions significantly reduces the chances of your observation being run. If your plan is long, there is a very good chance it will not get chosen for the robotic telescope due to time constraints or that it will not be submitted to the queue by the time allocation group, as it is an unrealistic use of resources. If you can cut your plan up into several smaller plans, then your chances of getting some or all the data are increased. If you are doing something that needs to run at a specific time, for example an exoplanet transit, talk to the time allocation group that allocates priorities, if there is one. They might be able to give the plan a high priority, since it can run in only one time slot, or they may offer remote mode operation instead.

On occasion, it is impossible to do something in the interface but possible in RTML or with remote operation. This normally happens because the telescope interface is designed to be easy to use and tends to cover the most common types of requests. Some telescope operators will let you submit raw RTML, but check with the observatory first and always be prepared to justify why you are not using their shiny interface on which they spent a small fortune.

If you are undertaking remote observation, then ensure that you are familiar with the observatory's operation, ideally by actually using the telescope in the dome. Make sure you have read all the documentation and have the emergency contact details for the observatory. If there is a minor problem, for example you are struggling to focus, then you might wish to contact the observatory staff if they are located on-site. If they are not, just send them an email, and they will fix the problem when they start work. More than likely, you will get the time back. If you have a major problem, like the dome not closing (and yes, that does happen), you should contact the emergency number. The telescope, and more likely the sensitive electronics, could be damaged if the telescope is left open to the elements. If you lose your connection and cannot reestablish it to force the dome to close, you should also contact the emergency number unless you can see the dome closing, for example via an in-dome webcam. Most observatories that have remote and robotic operation have some kind of emergency shutdown trigger that is actuated by the remote connection dropping out or by a trigger from a weather station on site. However, often they are not set up to deal with a loss of power, which is what might have happened if you have lost your connection, although normally that won't be the case. The best practice is to follow the guidelines in the documentation you are given and not to be too timid about asking for help.

Chapter 9
Image Reduction and Processing

Abstract Image processing is the process that turns an image from a science image to something more pleasing to the eye or highlights a structure of interest. Many image processing techniques render the image unsuitable for photometry. In this chapter, we discuss image processing techniques such as stacking, sigma clipping, drizzling, stretching, as well as the application of image processing filters. We touch on how to make colour images from black and white astronomical images using commercial packages.

9.1 Introduction

In this chapter, we will discuss the processes by which you take your raw light image or images and create an image that has been modified either for aesthetic or scientific reasons. Although these two goals are not always in opposition, most of the processing required for the production of high quality glossy astronomical images renders them all but useless for scientific purposes. Hence we should distinguish between **image reduction**, which we can consider the process required to turn a light frame into a science frame, and **image processing**, which goes beyond image reduction to produce an image that is pleasant for the viewer.

It is therefore important to understand your final goal. If you wish to perform photometry or astrometry, you will not be making coloured images, nor will you apply highly sophisticated mathematical filters to your image. However, if you wish to look at the structure of the object, then coloured images and filtering are useful tools. Likewise, if you are looking at producing spectacular images, then your stacking technique will be different than that employed for stacking photometric images. If you wish to process your image, you may wish to use processing software such as Adobe's Photoshop or The Gimp to tidy up an image.

Observational astrophysics rather than astronomical imaging is the principal subject of this book; therefore, this section concentrates mostly on image reduction rather than image processing. However, most image reduction and image processing techniques are the same. We will, however, touch on image processing and the pro-

© Springer Nature Switzerland AG 2020
M. Gallaway, *An Introduction to Observational Astrophysics*,
Undergraduate Lecture Notes in Physics,
https://doi.org/10.1007/978-3-030-43551-6_9

duction of colour images, although a full treatise on the matter is beyond the scope of this book.

The basic workflow of image reduction and processing is calibration, alignment, stacking, combining, and enhancement, the last two of which apply only to image processing. Combining is simply the process of turning a greyscale (i.e., black and white) image into a colour one. Enhancement covers a number of processes, from enhancing faint detail to making stars smaller and less dominant and even to adding diffraction spikes!

There are a number of general and astronomy-specific software packages for image reduction. Some, for example MaxImDL, are both image acquisition and image processing packages, while others, such as StarTools, are for image processing only. Packages such as Photoshop and GIMP are nonspecialist image processing software, and you will need some further software prior to processing your images in those packages. Some packages, for example DeepSkyStacked (DSS) and FIT Liberator, perform very specific reduction tasks, such as image format conversion.

9.1.1   Calibration

Calibration of images must take place before processing, i.e., the dark bias and flat frames must be applied. Figure 8.2 shows the pathway needed to create and apply those frames. If you intend to stack images, you must apply the calibration frames *before* stacking. As stated previously, most camera control software allows you to apply calibration frames (for example MaxImDL), as do many of the post-imaging packages, such as AstroImageJ and GIMP.[1]

The order of calibration is the subtraction of dark and bias frames from the light frame followed by the division of the resultant light frame by the normalised flat frame. The resultant frame is the science frame. If you have only single images, you may wish to perform hot and cold pixel removal at this point. Bad pixels are not so much of an issue if you intend to undertake frame stacking.

The calibrated image should be visually checked for quality. Checking the images might be time-consuming if there is a large number of them. However, stacking a poor image adds nothing to the overall image quality, and flaws such as vignetting and aircraft trails are difficult to remove except by exclusion. Images may be checked for quality before calibration. However, a poor or misapplied calibration frame might go undetected if images are not checked post calibration.

---

[1] GIMP requires a plugin in order to undertake calibration.

9.1.2 Stacking

Once the calibration frames are applied and you have the science frames, you will need to stack them if you have taken multiple science frames in the same band. In order to stack, you need to align the frames first. There are several packages that can do this, including MaxImDL, Deep Sky Stacker (DSS), and AstroImageJ; both DSS and MaxImDL will align and stack in what appears to be a single operation, although of course, it isn't. The easiest alignment method, and in my opinion the most successful, is to use the WCS in the header to align, although this is possible only if the frames are plate solved. Pattern alignment is used by both DSS and MaximDL. Essentially, an algorithm identifies the stars in the field and uses them to make a series of virtual boxes with stars at each corner. These *boxes* are then matched by rotation, translation, and scaling of the frames until they match. An alternative to this is an aperture or star matching. Here the user does the hard work by identifying the same group of stars within each image, the software using this data to find the required transformations. Lastly, some applications have the option of comet alignment. Comets and other solar system bodies tend to move against the background stars over multiple frames. Traditionally, stacking methods would stack the stars at the expense of the comet or another body. Comet stacking (or if this is not available, single star matching, whereby the solar system body is selected rather than a star) stacks to the body being observed rather than the background stars.

Once the frames are aligned, they need stacking. If this is going to be done during the alignment process, it is vital that the stacking process be selected *before* the alignment is started. There are several ways to stack a series of frames, collectively known as, not unsurprisingly, a stack. The process you use will depend on the frames, the target, and what operations are going to be undertaken after stacking.

The simplest form of stacking is **summing**. This process simply performs pixel to pixel addition that keeps the pixel information linear, and if stacked to a higher bit, FITS can be used to stack images with stars that would otherwise saturate. Summing is ideal when photometry is being undertaken. Ensure if photometry is intended that the exposure time is set to the sum of all the co-added frames. The software will likely do this, but it doesn't always.

Mean stacking, as the name implies, is the same as summing, but the final pixel value is divided by the number of frames. This process will result in a linear science frame if all the frames in the stack have the same exposure time. Mean stacking has a positive effect on bad pixels if the image is dithered, as the effect of the bad pixel is spread over several pixels.

Median stacking uses the median value of the pixel to pixel stack. Median stacking has similar results as mean stacking, although I feel that it handles artifacts, defects, and noise better. However, it does not preserve pixel linearity, so if you wish to do photometry, I suggest avoiding median stacking.

Sigma clipping, or kappa sigma clipping, determines the standard deviation over the pixel stack and removes pixels with values displaced more than a set number of standard deviations, kappa, away from the mean before reducing the pixel value to

Fig. 9.1 The stacked image of NGC281 (the Pacman). The stack consisted of 15,300 s hydrogen α frames. The stacks were aligned using the WCS and stacked using $\kappa = 3$ sigma clipping. (Image courtesy University of Hertfordshire)

the mean. Sigma clipping is therefore very effective at removing cosmic rays or bad pixels, as they tend to have outlying values on a single image, assuming that you have dithered. However, it produces a nonlinear pixel response (Fig. 9.1).

Drizzling is a stacking technique developed by Richard Hook and Andrew Frucher in order to improve resolution. In essence, the process stacks the images to a much finer grid than that imposed by the pixel layout of the CCD. This process should improve resolution for images that are undersampled, i.e., the pixel size is larger than the seeing.

9.2 Stretching the Curve

We have already encountered stretching in undertaking initial imaging. Stretching is the process of changing how pixel values map onto the display. Stretching is required because the full range of the pixel value far exceeds the human eye's ability to perceive small changes in brightness or a monitor's ability to display such changes. Effectively, we want to bin the pixel values. For example, most images are typically

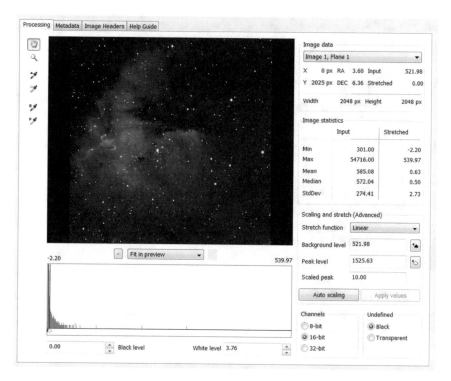

Fig. 9.2 Example of a manually stretched image, in this case, Fig. 9.1 stretched in FITS Liberator. (Image courtesy University of Hertfordshire and ESO)

8-bit, displaying brightness levels from 0 to 255. For colour displays, this would involve three or four values (red, blue, and green and sometimes a black level). Clearly, this is much smaller than the pixel range within your camera, so we might decide, as anything less than the mean pixel value is likely to be background, to set any pixel with a value less than the mean to a display value of zero. Likewise, anything near the high end of the pixel value is likely to be a star, so we might wish to set any pixel value in the top 10% of values to represent a display value of 255 (Fig. 9.2).

There are a number of ways we can perform a stretch. The simplest, and the one we have already encountered, is to stretch manually using the pixel value distribution histogram. The histogram shows the most common pixel values, with the peak typically being the background. By moving the upper and lower display limits, we reduce the pixel value range to which each display pixel corresponds, thereby enhancing detail. Stretching in this manner is known as a **linear stretch**.

An alternative method is to apply a built-in nonlinear stretch to the entire image, the most common of which is a **log stretch**. A log stretch enhances pixels of low value while reducing the impact of high-value pixels. In a log stretch, a pixel with a value of 10 would be translated to a value of 1, while one of 60,000 would be

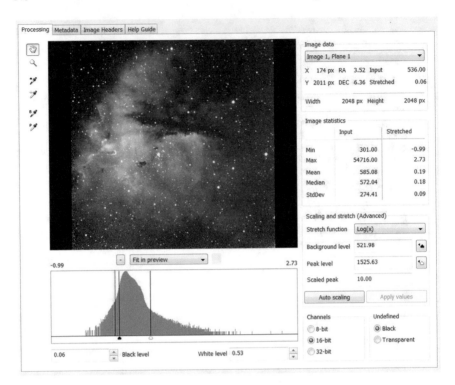

Fig. 9.3 Example of a log stretched image, in this case, Fig. 9.1 stretched in FITS Liberator. Note the difference in the pixel value histogram between this stretch and the linear stretch shown in Fig. 9.2. Also, note that there is a manual stretch on top of the log stretch. The entire histogram range is not displayed. (Image courtesy the University of Hertfordshire and ESO)

translated to 4.77 for a Log10 stretch. Other preset stretches include sin, loglog, and power law stretching. After a nonlinear stretch has been applied, it is normal to have to adjust the display range, i.e., to make a linear stretch on top of the nonlinear stretch, in order to make the most of the image (Fig. 9.3).

If you have both bright and faint structure, then linear and built-in nonlinear stretching are going to be unsuitable, and a curve stretch will be needed, which can be done both before and after the colour image is constructed.

MaxImDL, Photoshop, Gimp, and a number of other image processing packages have this function. The principle is that the relationship between pixel value and the display value is plotted as a line, and that line is manipulated in order to enhance the required details. Where the gradient of the line is flat, there is high compression of the pixel value, while where it is steep, there is low compression. Hence, the pixel values covered by the high gradient area of the curve are enhanced, and the ones in the low gradient area are suppressed.

For example, looking at the pixel value histogram for Fig. 9.1, we can see the main peak and a secondary peak with a pixel value around 3,800 counts. By a modified log

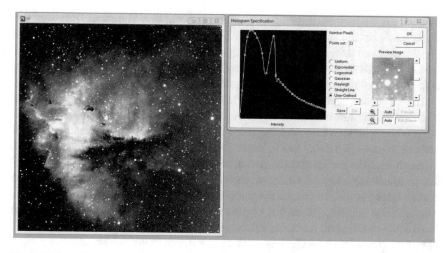

Fig. 9.4 Figure 9.4 shows the effect of applying a stretch curve to a FITS image. In this case, the image is Fig. 9.1 and the curve is applied in MaxImDL. (Image courtesy University of Hertfordshire and Diffraction Limited)

stretch using the curve option in MaxIm DL, both the pixels covered by the principal peak in the histogram and those in the secondary peak can be enhanced; the results of this process can be seen in Fig. 9.4.

9.2.1 From Greyscale to Colour

Once you have your individual filter images reduced and stretched, you may wish to combine them to make a colour image. Typically, you will make a colour image from three images, and assign a red, green, and blue channel to each. Normally, you would assign a filter to the channel that most represents that filter's passband. So, for example, an image taken with a Johnson V filter would be assigned to the green channel, while an I filter would be assigned to the red channel.

Of course, it is not always possible to simply match a filter to a colour channel, for example, for narrowband filters. There are conventions to help. For example, if you have V, B, and Hα, the Hα is assigned to the red channel. If you already have an R filter set, the Hα may be stacked with the R band image, and a four-colour image could be created. For the standard Hα, OIII, SII set there is a standard known as the **Hubble palette**, whereby SII is mapped to the red channel, Hα to green, and OIII to red.

The majority of astronomical image processing software will let you assign a frame to a channel, align those frames, and then turn them into an RGB image. With nonastronomical packages such as Gimp and Photoshop, you have to load the images as separate layers and assign colour channels to each layer, then align the channels by hand before combining them into a single colour image. You should be aware that Photoshop does not handle FITS files and GIMP does not always handle them well. If you are using an image processing package, I suggest therefore that you convert your files from FITS to an uncompressed standard format such as TIFF. FITS Liberator is the ideal tool to undertake this conversion and to perform your initial stretch.

In some cases, such as when a target has a lot of fine structure, it may be possible to add a fourth, **luminance**, channel. A luminance is a low signal-to-noise frame, often taken without a filter. Adding this as a luminance channel to create an LRGB image enhances fine detail lost in lower SNR filtered images. Many amateurs spend as much time on the luminance frame as on the rest of the image. If you do not intend to use the images for science, then a luminance is strongly recommended. If you do intend to use the images for science, the time spent on the luminance can be best used for improving the SNR of the science frame. You can imagine that in an RGB image each pixel has three values assigned to it, representing the mix of red, blue, and green. In an LRGB image, there is a fourth value, which tells the display how bright a pixel shall be. As the human eye is much more sensitive to changes in brightness than it is to colour, the effect of adding the luminance frame is to enhance the differences in brightness between regions.

Be aware that when processing LRGB images it is good practice to work on the combined RGB frame and the luminance frames separately in order to get the colour balance and stretching correct. You then break up the RGB frame into separate channels and then recombine everything into an LRGB.

There are almost as many approaches to creating an RGB (or LRGB if we have a luminance frame) as there are software packages. For example, MaximDL makes the transition from single images to colour image seamless in a one-step approach. Users select the image combination they wish, for example, RGB, LRGB, or CYMK (cyan, yellow, magenta and black for litho printing) and then assign a channel to each one. Once the data have been accepted, Maxim will create the image, which can be saved in most standard formats.

AstroImageJ takes a rather different approach. In AstroImageJ (which cannot handle luminance data yet), the individual frames are loaded into a stack. We can consider a stack to be a cube, with the x- and y-axes representing the pixel position and the z-axis the frame number, or slice number in AstroImageJ terminology. The user assigns a colour to each slice, and the cube is merged vertically to form a colour image.

Producing a colour image in Photoshop or GIMP is a little more complicated than using an astronomical imaging package. Hence, I will walk through the process for both applications.

I will start with Photoshop. I assume you will be using Photoshop CS6,[2] and we will be producing an LRGB image from four frames that have been previously stacked, aligned, and converted to TIFF or JPEG. Open your four frames and then add a fifth new frame. The setting for that frame should be the same as one of your open frames—it does not matter which. Set *Background Contents* to Background

---

[2]Although this instruction should work with most previous versions.

Color and *Color Mode* to RGB Color. Select the new image and the channel tab. You will see a list of four images, RGB, Red, Green, and Blue.

Select the red channel and then select the red image. Select the whole of the red image with a ctrl-A and then copy it with a ctrl-C (or just use the Edit drop-down menu). Now return to the new image and select *paste in place* from the Edit menu. Do the same for the green and blue channels until you have your colour image. You now need to add the luminance frame. Select the Layers tab, then Layer New Layer. Set the new layer mode to Luminosity. This will bring up a second layer above your RGB image (which will be called Background). Select the new layer and then select your luminance frame. Select all and then copy, then return to the new image and paste in place. You now have an LRGB image.

The process is somewhat similar in GIMP. Open your three RGB frames as we would in Photoshop, except we do not need to open the luminance frame yet. Also, we do not need a fifth frame. From Colors select component and then compose. Choose RGB as the Color Model and select the three frames corresponding to the three filter colours; pressing OK results in a colour image. To add the luminance frame, select the colour image and then select the luminance file from File Open as Layers. Highlight the new Luminance layer in the Layer bar and change its mode to value. Now left click on the luminance layer and select merge down. You now have an LRGB image.

9.2.2 Image Filtering

Now that you have your colour image, you may wish to enhance it in order to bring out the finer detail. This can be done by applying any number of mathematical processes to the image, a process known as, somewhat confusingly, filtering. Filtering an image is fundamentally different from stretching an image. Filtering is nonlinear, and once accepted, it is permanent. It does not change how the image is displayed, but it changes the pixel values. Hence, once you filter an image, you can no longer use it for the purposes of science. However, for completeness, we will detail a number of the more common image processing filters.

- **High Pass** We can consider the data in an image to be a series of waves representing the rate of change between pixels. If the values change rapidly between pixels, the data is of high frequency. The high pass filter accentuates the high-frequency regions of the image, enhancing those areas at the cost of increased noise.
- **Unsharp Mask** The unsharp mask performs edge detection and enhances the edges. In most implementations, you need to define the width of the edges, the aggressiveness of the process, and the floor at which a pixel is considered part of the background.
- **Wavelets** The wavelet process identifies specific frequencies (as selected by the user) within the image, and either enhances or suppresses them. The process is very powerful, but it requires considerable skill to use.

- **Deconvolution** Deconvolution is a complex mathematical process that deblurs the image by modelling the blurring from the point spread function (PSF). If there is no blurring in the original image, deconvolution will have no effect.
- **Erosion** Erosion filters remove bright edges around stars, making them both sharper and smaller.
- **Kernel Filters** A kernel filter applies a kernel, convolution matrix, or mask to the image. This is basically a matrix. The matrix is moved over the image, which is multiplied by the values with the matrix. The effect of the kernel filter is dependent on the contents of the matrix. It may blur, sharpen emboss, or enhance edges, depending on its parameters.
- **FFT Filters** A fast Fourier filter applies a Fourier transform to the image (recall that an image can be treated as a series of waves). FFT filters are used to reduce patterning on an image such as stripping.
- **Digital Development Filters** Digital development filters are a series of mathematical filters designed to make CCD images have a film-like response, i.e., nonlinear. They bring out faint detail and reduce bright sources.

Chapter 10
Photometry

Abstract Photometry is one of the key observation pillars of astrophysics. New techniques have improved the precision of photometry to the point that exoplanet transit observations are possible with small telescopes. This chapter details the methods by example to perform photometry, the reduction of images taken for photometric purposes, and the interpretation of the results.

10.1 Introduction

Photometry, the measurement of the amount of electromagnetic radiation received from an object, is one of the cornerstones of modern astrophysics. The ratio of two fluxes, in astronomy a colour, is an important tool for classification and identification of stellar objects.

Time-dependent photometry is used to understand the physics of many astronomical objects such as cataclysmic variables, accretion discs, asteroids and Cepheid variables. Cepheid variables are an important component of the distance ladder that leads to the measurement of the most distant objects in the observable universe.

High-precision, time-dependent photometry is one of the principal methods for the detection of planets surrounding stars, **exoplanets**, used in astrophysics.

The advent of the CCD, coupled with increased computer power, has considerably improved the accuracy, precision, and ease of photometric measurements. This coupled with modern small telescopes, such as those found in many university teaching observatories, means that students are now able to perform milli-magnitude measurements. Such precision is needed to observe and detect exoplanet transit, and once again, small telescopes are making a contribution to science and our understanding of the universe.

There are two fundamental approaches to photometry: aperture and PSF. And there are three approaches to the calibration of the results: absolute, differential, and instrumental.

When we measure the light from a distant object, we measure the light coming not only from the object, but also from the sky, along with stray interstellar light and scintillation noise, in addition to the false signal caused by the remnants of the bias

© Springer Nature Switzerland AG 2020

M. Gallaway, *An Introduction to Observational Astrophysics*,
Undergraduate Lecture Notes in Physics,
https://doi.org/10.1007/978-3-030-43551-6_10

and dark currents. These sources are collectively known as the **background**. The method by which we deal with the background is the difference between aperture and PSF photometry.

In **aperture photometry**, we drop an aperture, which is normally a circle, but not always, onto the image of the target star. We then integrate the count constrained within that circle. The diameter of this circle is important. The circle should contain only pixels associated with that star and be wide enough to capture the vast majority of the signal from the target. The circle's diameter typically has a radius three times the FWHM PSF of the star.

Recalling that the PSF should be Gaussian, you should see that an aperture of this size contains 99.7% of the light from the star. Because the PSF is principally a function of seeing, the aperture size should be constant for the whole field. The difference between a bright star and a dim one on the image is not the width of the PDF but the height.

Clearly, the integrated count within the aperture also includes a contribution from the background. It is impossible to determine directly which are from the source and which are from the background, so we must assume that the background is constant at local levels. By dropping an annulus onto the aperture, where the inner ring of the annulus is beyond the aperture, we can address the problem of a noisy background. Typically, the area contained within the annulus is the same as that enclosed by the annulus, but if not, it should always be larger. We can take the background to be the integrated count within the annulus normalised for the area of the aperture. Hence, stars falling into the annulus must also be avoided, as they would artificially heighten the background. We subtract this area's normalised background from our integrated aperture count in order to get an instrumental count. As the count will depend on the exposure time, we need to divide the count by the exposure time to get counts per second.

The key drawback of aperture photometry is that it requires that an aperture and annulus, without contamination from other stars, be fitted to the target. In very crowded fields, such as those found in globular and open clusters, this clearly is not possible.

The solution to this problem is PSF fitting. Given that a star's profile is Gaussian and a Gaussian is determined by its width and height, it is possible to estimate the count by fitting a Gaussian to the PSF, whence PSF photometry. In this way, even if the PSFs of two stars overlap, we can achieve a reasonable estimate of the count. In most cases of PSF photometry, the PSF is used to detect and remove the stellar contribution to the entire image count, and the remainder is used to determine the background.

In general, PSF photometry is not as accurate as aperture photometry in ideal conditions, but it outperforms aperture photometry in crowded fields. Many applications used for photometry can do both PSF and aperture photometry, and you should judge which one you should use based on the field and the FWHM of the stars in the image. You may have thought about the implications of unresolved binaries on photometry, two or more stars being so close that they appear as one. Fortunately, as luminosity increases by approximately $M^{3.5}$, low mass companions contribute little to the overall luminosity of the system. Even in a system comprising a pair of stars of equal mass, the colour is not overly affected, although any distance based on the measured brightness of such a system would be less accurate than a single system.

There are two approaches to calibrating raw **instrumental magnitudes**. The first of these is differential photometry, which uses reference stars, preferably of the same spectral class and in the same field of view and hence imaged at the same air mass.

Differential photometry is often used to determine the magnitude of variable stars and to identify exoplanet transits. The magnitude of the reference star does not need to have been accurately determined (as long as it isn't variable itself).

The measurement needed in differential photometry is the difference in magnitude between two or more sources, δMag, rather than an apparent magnitude.

Results of very high accuracy and precision can be obtained using this method. Determining an object's apparent magnitude differentially to anything better than a few tenths of a magnitude is challenging, mostly due to the dearth of stars with very well determined magnitudes. Both the Landolt (UBVRI) and Henden (BV) catalogues provide well-determined standard stars. However, these are often not accessible for observers at high latitudes.

The second method, **all-sky photometry**, uses a well-calibrated optical system to determine magnitude directly from images. All-sky photometry is used mostly for large-scale surveys and also for regions well away from a suitable calibration star. Photometric accuracies of a few hundredths of a magnitude are possible for a well-determined system, which is a factor of 10 poorer than that achieved by differential photometry.

In the rest of this chapter, I will concentrate on differential aperture photometry, as this is the form most commonly encountered by undergraduates. Differential aperture photometry is versatile, and many of its aspects form the foundations of other forms of photometry.

10.2 Measuring

In this section, I will walk through how to perform standard aperture photometry that will yield an uncalibrated instrumental magnitude using the **Aperture Photometry Tool (APT)**. APT is freeware, and as it is written in Java, it will run on almost any platform. Later on, I will be using a similar programme, AstroImageJ, to perform differential photometry. Like APT, AstroImageJ is freeware and multi-platform. APT is more suited to field photometry, measuring all the stars in a frame, while AstroImageJ is more suitable for time series photometry, measuring just a few objects but over many frames.

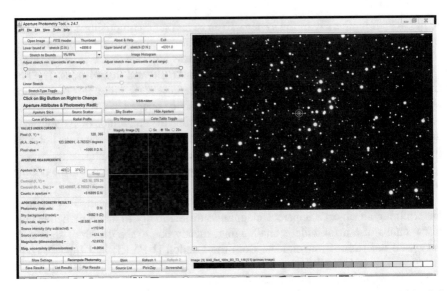

Fig. 10.1 Screen shot of aperture photometry tool with an image of the open cluster M48 loaded and a star selected with an aperture and annulus overlaid. (Image courtesy of the University of Hertfordshire)

10.2.1 Aperture Photometry

Before we begin the photometry, I suggest checking the FITS header of your image to determine whether it has been plate solved. If the header contains cards with the heading CTYPE1, PLTSOLVD, or PA, which are WCS keywords, then it has been plate solved. This will make things a lot easier later on. If it has not, you can either use a local plate solving service, which some applications have, or you can push the image up to the astrometry.net website for solving. It is not necessary to do this, but in my experience it is highly desirable.

Once an image is open in APT, we need set the parameters for our photometry, which can be found in the more setting button. We select Model 2 and Model D. This gives us sky subtraction and takes into account that some pixels are intercepted by our apertures.[1] When you now left click on a star, a series of concentric circles is displayed that comprise the aperture and the annulus (see Fig. 10.1). Most likely these are set incorrectly. We need to know the FWHM of the frame in order to set it correctly. If you have used an autofocuser or the image is plate solved, you might find the FWHM listed in the header, which should be in pixels (but always check). If not, you will need to plot the PSF of a field star. In ATP, this is done by selecting a field star and pressing the Aperture Slice button.

In ATP, this displays both an x and a y slice through the PSF of the selected star. The star might not be completely circular, so the two slices may not match, in which

---

[1] Do not forget to press the apply button!

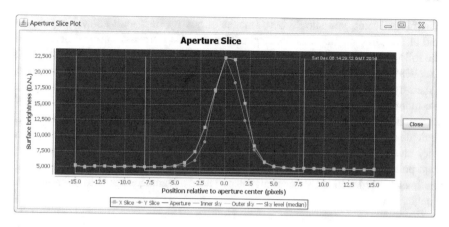

Fig. 10.2 Screen shot of aperture photometry tool aperture slice showing the PSF of the star selected in Fig. 10.1

case just select the widest of the two. In the case of Fig. 10.2, half the count of the peak is about 14,000, as we need to subtract the background before we halve the height and then add it back. The full width at this point is about five pixels (which happens also to agree with the number in the header). Good practice is to set the diameter of the aperture to three times this number, that is, 15 pixels.

The radius of your aperture is set under the Big button, which raises a popup with a direct entry box and a slider used to set the aperture radius. The multiplier radio buttons (1×, 5×, 10×, and 20×) set the size of the annulus based on the aperture size. As the aperture radius is an integer, I would set this to 8, rather than 7. Note that unchecking Center(x, y) lets you make elliptical apertures, which are useful in extragalactic photometry.

When you click on the stars in the image window of the Aperture Photometry Result Panel, you might notice that the magnitudes look strange, often negative. The magnitude is negative because it is an instrumental magnitude, and you will have to calibrate your data into apparent magnitude later.

Although APT reported the magnitude, it is the source intensity (sky subtracted) and the sky scale source uncertainty that we wish to record, along with the RA and Dec of the object. Be aware that some software time normalises the results.

10.2.2 Photometric Reductions

If you need photometry in more than one filter, you will have to undertake some form of photometric reduction. These reductions turn your counts into reduced standard magnitudes accounting for the instrumentation you are using, under the conditions and the air mass at which the observations took place. For most purposes, there should be stars within the field with known magnitudes in the bands in which you are observing. If not, you should have images of nearby stars with known magnitudes and at the same air mass as the target. The reduction process outlined here and in the following practical is discussed in considerably more detail in Brian D. Warner's excellent book *A Practical Guide to Lightcurve Photometry and Analysis*, also published by Springer. There are techniques that claim better accuracy, but the gains are marginal given the complex nature of photometric reductions and the inherent uncertainties in small telescope systems, which are often based at sites that are less than ideal. When imaging for photometry purposes, it is best to take multiple images very close together in time. The use of multiple images increases the number of data points in your photometry, thereby reducing uncertainty and ensuring that all the images are taken at the same air mass. These advantages apply to reference star images as well.

Previously in this chapter, I discussed how to set an aperture size and measure the integrated count within the aperture and how to account for the sky. Your software should, unless it's very simple, like DS9, be able to automatically time normalise the data. If it doesn't, do not forget to divide the background subtracted count by the exposure time in seconds. You should determine the instrumental integrated background subtracted count for all the reference stars you are using and for every image. Take the mean count value for each star and use this as the instrumental count (and also calculate uncertainty from this). The mean instrumental count needs to be changed to an instrumental magnitude using a modified version of Pogson:

$$m = -2.5 \log(count), \tag{10.1}$$

where m is the raw instrumental magnitude, and count is the count within the aperture minus the background.

The count does not represent the number of photons received; rather, it is a ratio between the electrons generated and the **gain** of the camera. You will find the camera gain in the camera manual or written into the FITS header as EGAIN. Note that gain might also be determined by the binning. The gain is applied to the count before it is logged. Hence, (10.1) becomes

$$m = -2.5 \log(count \times gain), \tag{10.2}$$

and your errors become

$$\delta(m) = -2.5 \log 1 - \frac{count \times Gain}{\delta count \times Gain}, \tag{10.3}$$

where $\delta flux$ is your calculated count error.

If you now perform aperture photometry on a star in a science frame and apply (10.2), you will find that you obtain an odd-looking result, most likely a magnitude with a very negative value. The reason for this is that your photometry is not reduced. The process of reduction takes your raw instrumental magnitude and converts it into a standard magnitude, one set up for observing above the atmosphere.

The basic reduction formula in magnitudes is

$$M_f = m_f - k'_f X + T_f(CI) + ZP_f, \tag{10.4}$$

where M_f is the standard catalogue magnitude of the reference air massstar in band f, m_f is the instrumental magnitude of the reference star in band f, k'_f is the extinction coefficient in band f, X is the air mass, T_f is the colour correction for band f, CI is the colour index, and ZP_f is the current band f zero point.

The **zero point** is an offset between the instrumental magnitude and the standard magnitude. In broad terms, it is a measure of the sensitivity of your optical system to light. Hence, space-based telescopes also have a zero point. A zero point can be either a flux or a magnitude. In the case of a flux, it is applied by dividing the time and gain as a normalised count by the zero point. If it is expressed as a magnitude, it is just added to the instrumental magnitude, as shown in (10.4).

The **extinction coefficient** reflects the proportion of light that is lost as it travels through the atmosphere for a given air mass and hence is both wavelength and filter dependent, as well as being altitude specific. The **colour index** is the ratio of flux received from a star over two bands, or more understandably, the difference in magnitude between two bands for that star. So, for example, the star HIP 2027 has an R magnitude of 7.574 and a V magnitude of 7.640 and hence a V-R colour index of 0.065.

Different wavelengths are subjected to differing amounts of extinction as they pass through the atmosphere, with more blue light being scattered than red (hence the sky appearing blue). A blue star will therefore be affected by the atmosphere more than a red one, whence the need to apply an extinction coefficient that is air mass and filter specific. Note that the air mass is a function of the altitude at which the object is observed, and applying

$$X = 1/\cos(z), \tag{10.5}$$

where $z = 90 \deg -altitude$ gives the air mass X. Note that this is the altitude angle of the object, not the altitude of the observer.

The colour transformation $T_f(CI)$ is effectively stable and can be used on successive nights unless there have been significant changes in the optical system, and likewise for the extinction coefficient. The zero point is related to the entire optical pathway and will change between individual nights of observing, so it should be recalculated for every session. The air mass X is a function of the actual observation.

You may have heard the terms first- and second-order extinctions in the context of photometric reductions. The first order applies to the air mass coefficient, whilst second order applies to the colour transform. In some cases, the first-order extinction can be ignored if you are using calibration stars at the same air mass. However, as this might not be the case and given that they are largely immutable, it is worth having their values at hand.

The following practical will guide you through the process needed to identify the individual components of (10.4). Once this is completed, you will only need to identify the zero point during subsequent observations while the optical system is unchanged.

10.3 Practical: Find the Zero Point and Transformations

Although raw, instrumental magnitudes are useful if you wish to undertake observations that may span more than one night or that require colour information, you will need to reduce your photometric data so that observations between nights or objects of different colours or observations at differing air masses can be compared. In order to transform instrumental magnitudes into standard magnitudes, you need to find your system's extinction coefficient, colour transformation, and zero point.

This practical draws on many of the skills you will have obtained in previous practicals, especially the Messier marathon practical, the bias and dark frame practical, and the flat frame practical. If you have not done these, it is recommended that you read them before attempting this practical.

10.3.1 Aims

In this practical you will find the nightly zero point of an image as well as the colour transformation and extinction coefficient for your optical system. You will also need to determine levels of uncertainty in standard magnitude when these parameters are used to convert instrumental magnitudes.

10.3.2 Preparation

In order to achieve the aims of the practical you will need to image three fields in at least two photometric bands, preferentially B, V, and R at a single binning value. These fields are the first-order field, the second-order field, and the uncertainty field. The fields you will need to observe should contain a number of photometric standard stars. Ideally, these will be Henden, Hipparcos red–blue pairs, or SDSS blue–red pairs (lists of these are carried in *Light Curve Photometry and Analysis*, Warner 2007). Your second-order field will need to be imaged as it crosses the local meridian at least three times, without overexposing the photometric standards. The first-order fields will need to be imaged a number of times at differing air masses. Again I suggest a minimum of three times per air mass and a minimum of three air masses. Again you should ensure that you do not overexpose the photometric standards.

You will need an uncertainty field, again containing photometric standards, that will be used to determine the uncertainty in your photometric calibrations. The calibrations should be imaged close to the meridian but not at the same air mass as either of your other two fields.

Once you have identified your fields, you will need to produce finder charts with the photometric standard marked with their magnitudes and the time that you should start and stop imaging with the exposure time for each band.

Unlike most of the other practicals in this book, this practical requires very good photometric conditions. There can be no Moon at the time of imaging, and the weather must be very good, with no cirrus. Good seeing is desirable but not as important as the Moon status or weather.

10.3.3 Imaging

When you arrive at the observatory, get the camera down to its operational temperature as usual and ensure that it is ready for use. You will need to take multiple flat, bias, and dark calibration frames in order to make master calibration frames. This practical will suffer if you are not careful with your calibration frames, especially the flat frames.

1. As usual, before you start science imaging, make sure the telescope is focused, that the binning is set correctly, the autoguider is operating correctly, and that the camera is not subframing.
2. Slew the telescope to the first field some time before you intend to image and take a point image to ensure that the telescope is pointing at the correct field and is centred.
3. As the target reaches the required altitude, start imaging. Take the images as a filter sequence, i.e., VRBVRBVRB rather than VVVRRRBBB.
4. You will have some time between image sets. Use this time to check that the images are of acceptable quality, with no trailing, overexposed stars, or plane trails.
5. Once you have imaged all your first- and second-order fields together with the uncertainty field, shut the telescope down (if required). Ensure that you have all the data on a removable medium, including the calibration files.

10.3.4 Photometry

1. In order to perform photometry, you will need to turn your light frames into science frames, as outlined in Chap. 9.
2. Once you have your science frames, you will need to find the seeing of each image. If you are using APT, this can be done by dropping an aperture of a star

and taking an aperture slice from which you can get the FWHM. In MaxImDL, you need to use the Graph function (found under View. Select Line and draw a line through a star. Doing so will give you a slice of the PSF, from which you can obtain the FWHM. In AstroImageJ, this is done by drawing a line through a star and performing a Profile Plot (found under Analyse). In theory, the FWHM should be the same for all of your images. If it is significantly different, then do not use the outlying images.

3. You now need to set the aperture and annulus size. Set this to three times half the FWHM. In general, it is best to let the software determine the size of the annulus, but if it does not, five times the aperture radius is normally suitable. If you have a very dense cluster, you may wish to have a much smaller annulus size in order to avoid contamination from nearby stars.

4. Using ATP, you will need to set the source algorithm to Model 2 and the Sky algorithm to Model D.

5. Drop an aperture on each of your reference stars, noting the source count and the count uncertainty. In APT, this is done by double clicking on the target box. APT uses a very sophisticated uncertainty model that goes beyond a simple signal-to-noise ratio. If you are not using APT to do the photometry, use the SNR to calculate the uncertainty instead.

6. Using the camera gain, which should be in the FITS header, time normalise the count, if needed, and apply (10.2). This will give an instrumental magnitude for each reference star in all fields.

7. Record each instrumental magnitude along with the target star's position, the filter in which the image was taken, the air mass, and the magnitude of the standard star.

10.3.5 Reduction

There now follows a series of steps that will find the various components of (10.4) for your system. These processes require the plotting of data and the fitting of a line to that data. In Warner, it is suggested that a spreadsheet be used. You could use a specifically designed application, such as PhotoRed, or perhaps a mathematical application such as Matlab to plot the data and fit the line. However, whatever you use, you should always visualise the data. It is easy for outliers to change the straight line fitting dramatically. If you see an outlier, try to determine what caused it. It might be a hot pixel, a cosmic ray, or even a misidentified reference star, and address the problem if possible. If you cannot correct it, you can remove it from your dataset as long as you make a note of it in your lab book, with an explanation of why it was removed.

In this section, I will be assuming that you are using V and R filters for notation, with V and R being the standard magnitudes of the reference stars. The notations v and r are used for the instrumental magnitudes, and $\langle v \rangle$ and $\langle r \rangle$ for the mean instrumental magnitudes. If you are not using V and R, just substitute.

Second-Order Extinction and the Zero Point

In the context of a V filter observation, we can rearrange (10.4) to

$$V = v - k'_v X + T_v(CI) + ZP_v. \tag{10.6}$$

As the second-order fields are all imaged at the same air mass we can safely ignore the first-order part of (10.6), reducing it to

$$V - v = T_v(CI) + ZP_v. \tag{10.7}$$

You should be able to see that (10.7) is the equation of a straight line with $T_v(CI)$ the gradient of the line and ZP_v the intercept.

You need to find the mean values for your v band and r band instrumental magnitudes ($\langle v \rangle$ and $\langle r \rangle$ respectively) for each standard star in the field. Plotting $V - \langle v \rangle$ against $V - R$ and fitting a straight line to this plot yields the V band colour transform T_v as the gradient and the V band zero point ZP_v. Performing the same plot with $R - \langle r \rangle$ against $V - R$ produces the colour transform and the zero point for the R band.

Once this is done, you need to find the **hidden transformations**. These are effectively adjustments from your instrumental setup's colour index to the standard one. They are hidden because they do not appear in (10.4). They are found by plotting $V - R$ against $\langle v \rangle - \langle r \rangle$ and fitting a straight line to give T_{vr} as the gradient and ZP_{vr} as the intercept. Undertaking the same process for $R - V$ against $\langle r \rangle - \langle v \rangle$ gives T_{rv} as the gradient and ZP_{rv} as the intercept.

First-Order Extinction

We now move on to our first-order field. We know that the colour transformation and the zero point are air mass independent and that they are constant during a night's observation. So, if we vary the air mass at which we observe calibration stars, the change in instrumental magnitude when adjusted for colour transformation and zero point is due to the change in X. Hence, we can find the extinction coefficient.

In order to determine K'_v, we need to plot the adjusted mean instrumental magnitude $\langle v \rangle_{adj}$ against air mass: $\langle v \rangle_{adj}$ and $\langle r \rangle_{adj}$ are found by applying

$$\langle v \rangle_{adj} = V - v - (T_v \times V - R). \tag{10.8}$$

Plotting $\langle v \rangle_{adj}$ against X and fitting a straight line gives the first-order extinction coefficient k'_v as the gradient, which should always be positive. It should also always be greater than or equal to k'_r. The interception point should be very close to the zero point of the filter in question.

Error Analysis

You will now have to look at the uncertainties in your magnitude calibration system. Because the steps involved are complex, the easiest way of determining the uncertainty is by comparison with known magnitudes.

Using the instrumental magnitude for each of your calibration stars in your uncertainty field, calculate the instrumental colour index for each calibration star using

$$CI_i = (((v - k'_v \times X_v) - (r - k'_r \times X_r)) \times T_{vr} + ZP_{vr}, \qquad (10.9)$$

where X is the air mass for that image and filter. Use the mean value for each star observed for the star's instrumental colour index, CI_i.

You can now transform the instrumental magnitudes to standard magnitudes using the mean instrumental magnitudes for each calibration star and applying

$$M_v = \langle v \rangle - (k'_v \times X) + (T_v \times CI_i) + ZP_v. \qquad (10.10)$$

By looking at $M_v - V$ you will be able to determine the level of uncertainty in your photometry and see whether there are any systemic errors.

10.3.6 Analysis

As usual, you should write your analysis in a standard format including all the data, plots, and calculations. Give special attention to discussing the level of your uncertainties given that a typical survey has photometric uncertainty errors of 0.1 mags. Where do you think your errors are and how do you think you could improve them?

In your discussion, include how you would be able to determine just the zero point if you were observing on another night. Observers of exoplanet transits might need photometric uncertainties of 0.001 mags. How do you think they can achieve this given the levels of uncertainty in your results?

10.4 Differential Photometry

One of the questions asked in Practical 8 is how exoplanet transits can be detected if millimagnitude precision is needed to give the level of uncertainty in our photometry as determined in the practical. The answer lies in the second-order extinction coefficient determination. When we image the reference stars in the same field as the object of interest, we can ignore the first-order coefficients. For most observations, we are not interested in the change of colour of an object over time, although that itself is interesting and contains a considerable amount of physics. Rather, we are interested in the change in magnitude ΔM in a single band. As both the colour transform and the zero points are time independent, these can be ignored under these circumstances. Hence, in this case, we are interested only in the difference between the instrumental magnitude of the target and the reference, although in general, multiple reference stars are used in order to reduce errors. Obtaining millimagnitude uncertainties using differential photometry is possible using a small telescope. However, you should be aware of the sources of errors so they can be minimised.

There are a number of sources that contribute to system noise with regard to photometry. **Read noise** (σ_{read}) is simply the noise generated by the electronics during the read process. It is effectively the noise within the bias frame (although not the bias current, which is systemic). The bias will vary across the chip and to some extent between bias frames. In general, the bias noise is small, with much of the variation taken out using a master bias, i.e., the mean of a number of biases. However, using a combination dark and bias frame (remember that a dark has a bias frame inside unless it is removed) is generally acceptable for imaging, although it does increase uncertainties in photometric measurements and should be avoided when photometry is being undertaken.

As discussed in Sect. 7.1, the dark current is the build-up of thermal electrons within the CCD. It is effectively, but incompletely, removed by the subtraction of the dark frame. So it is another source of random uncertainty (σ_{dark}). Again master dark frames are used to reduce dark current noise. Remember, however, that a dark frame includes a bias. Although the dark current will not vary across the frame as much as the read noise, it is, unlike the read noise, dependent on the exposure time and operating temperature.

Flat fielding, as discussed previously, is a challenging issue. For bright objects, the quality of a flat field and the noise of the flat field (**flat field noise**) normally are not significant. However, for a faint source or when measuring sources in multiple locations in the frame, it can become significant, because it causes variation across the field due to the field not being entirely flat. Therefore, a variation in pixel location can result in a differential pixel response.

In general, when performing single target photometry, the best practice is to keep the target in the centre of the frame, as this is where the frame is flattest. However, when undertaking differential photometry, this might not be possible for both reference star and target. We might require moving between the source and the reference, **nodding**, so that each appears on the same pixels, thereby reducing flat field effects.

As stated previously, because of atmospheric effects the light from a star does not appear as a point source occupying one pixel, but rather a circle (ideally) occupying multiple pixels. However, pixels are square and are spaced with gaps between them, and a small amount of light will fall between them, and the edges of the star will cut through pixels rather than completely fill them. This leads to **interpolation errors**. For the most part, the software will deal with this problem, but you should be aware of the problem if, for example, you are using DS9.

Another photometric precision problem is scintillation noise. Small-scale turbulence in the atmosphere Effectively scatters the light, causing the random nondetection of photons. Scintillation occurs on small scales of about 30 cm and over time scales often less than a second. Hence, it can be reduced with larger-aperture telescopes or increased exposure time. It is interesting to note that for larger telescopes, increased exposure time is often not an option, since it would lead to saturation.

Our last source of noise is **sky noise**. Random scattering, absorption, and emission of photons cause sky noise. It should be the largest contributor to the uncertainties and is the dominant factor limiting observable magnitudes. Fundamentally, without moving to a very high site, preferably in space, sky noise is going to be the principal limit on photometric precision.

It should be clear that most of these sources of error, besides interpolation errors, form part of the background of the science frame. Hence, the best way to reduce these errors' impact is to improve the signal-to-noise ratio.

10.5   Practical: Constructing an HR Diagram of an Open Cluster

The Hertzsprung–Russell (HR) diagram embodies one of the most important concepts in modern astrophysics. Its construction tells us about the evolution of stars and can be used to determine the age of a star cluster by the identification of the main sequence turn-off. Constructing and interpreting a Hertzsprung–Russell diagram is a vital skill for any undergraduate astrophysicist.

10.5.1   Aims

The aim of this practical is to construct an HR diagram and identify the main sequence turn-off. The construction of the HR diagram will require taking and reducing images of an open cluster in two bands. In addition, you will need to undertake aperture photometry on both images and plot the results. You should also prepare a backup cluster.

10.5.2   Observing

Once at the observatory, you should follow the standard imaging procedures:

- Cool the camera down to temperature and open the dome.
- Once the camera is down to temperature, take flats if needed.
- Slew the telescope to a focusing and alignment star. Confirm alignment and focus the instrument.

- Slew to the target. Take a short exposure image to confirm that you are on target and still in focus.
- Turn on tracking and select the first filter. Take several images of the target at different exposure times until you are happy that you have a good linear image.
- Undertake the same procedure for another filter.
- Take bias and flat frames.
- Save files to a portable medium and shut down the observatory.

10.5.3 Reduction

Once you have the images, you need to reduce them, perform photometry, and plot the results.

- Produce science frames by subtracting the bias and dark frames from your light frames and dividing by the flat.
- Open the first science frame in ATP (or a suitable alternative).
- Select a star and perform an aperture slice. Use the FWHM of the slice to set the aperture and annulus radii. Set the sky algorithm to median sky subtraction.
- Select a source list and then create a new source list. Process the source list.
- Hitting the List results button will display your photometry results. Export this file and then clear the contents.
- Load the second image and repeat the photometry process listed above.
- Load both photometric results files into a spreadsheet. Move all but the centroid RA, centroid Dec, magnitude, and magnitude error columns.
- If you know the filter's zero points, apply this information to the magnitude columns. This is not required if you do not have them.
- Crossmatch the two source lists (for example in TOPCAT) so that you have both magnitudes for most stars in the field (some will always be missing a magnitude, in which case discard them).
- Plot magnitude in filter A minus magnitude in filter B versus magnitude in filter.

10.5.4 Analysis

Write up your report in the standard format. You should include all your plots and observational logs as usual. You should discuss your errors and their implications. You should identify the main sequence in your plot and the main sequence turn-off. Discuss whether all the stars you imaged are cluster members and the implications if they are not. What do you think is the benefit of imaging a cluster, rather than a normal star field?

Chapter 11
Errors

Abstract All measurements, no matter how carefully undertaken, have errors. These errors need to be quantified so that different experiments can be compared. Results may have precision, self-consistency, and/or accuracy close to the real result. Results may have random errors and/or systemic errors to be addressed. This chapter explains the techniques for calculating and minimising the errors within an experiment as well as methods to propagate the errors when performing mathematical operations on the results.

11.1 Introduction

You will often see scientific results quoted as $x \pm y$, where y indicates the range of possible error. Errors, also called perhaps more accurately uncertainties, are an expression of how much confidence can be put in a result. It does not mean that a mistake has been made or that the scientists think that their result is wrong in some way. In fact, all measurements are to some degree affected by uncertainness, because a level of randomness is a fundamental part of the universe as expressed by both quantum mechanics and chaos theory.

However, understanding how much of your result consists of random events is an important skill, and you will be expected to express uncertainties for all your experimental results. Not only do the uncertainties express a level of confidence, they also allow comparison between different experiments. Say, for example, that one person does an experiment that gives a result of 2.5 ± 0.1, while another person employs a different method to make the same measurement and gets 2.6 ± 0.2. Although the quoted results differ in value, they are experimentally consistent due to their uncertainties.

The terms accuracy and precision are much more constrained in science than in everyday life, where they are used interchangeably. The term accuracy reflects how close a measurement is to its real value, while precision refers to how self-consistent the results are. It is possible to be precise without being accurate and accurate without being precise. For example, later on in this book, you will be asked

© Springer Nature Switzerland AG 2020

M. Gallaway, *An Introduction to Observational Astrophysics,*
Undergraduate Lecture Notes in Physics,
https://doi.org/10.1007/978-3-030-43551-6_11

to undertake differential photometry, in which you compare two stars, one with a known magnitude, and measure the difference in brightness. If the magnitude of your known stars were one magnitude out, your observations could be very precise but not accurate. When you use the words accuracy and precision, use them correctly and with care; it is a very common mistake to use them as if they are interchangeable—they are not.

Uncertainties can take two forms: random and systematic. Random uncertainties are caused by random and unquantifiable effects, for example cosmic rays and sky brightness. Systematic errors, on the other hand, cause a constant but unknown shift in the data; for example, it might be the bias on a CCD or to some extent the dark current. We don't know what the actual value is, but we can allow for it.

11.2 Decimal Places and Significant Figures

Another common mistake is to confuse significant figures with decimal places. The number of significant figures is a measure of precision, while the number of decimal places reflects accuracy. Hence, 11.230 is a result to five significant figures but three decimal places, while 0.001 is to three decimal places but only one significant figure.

When analysing your data, you should use only a level of accuracy that reflects your equipment and your ability to read it. For example, the count of a pixel is always an integer; hence to talk about a fraction of a count is meaningless.

The are several rules for dealing with significant figures and decimal places. For significant figures, all nonzero numbers are significant, so 12.23 is significant to four figures. A zero between nonzero digits is also significant, as are zeros after a decimal point; hence 101.25 and 101.00 each have five significant figures. Zeros between a decimal point and a nonzero number when the whole number is less than one are not significant. Hence, 1.002 has four significant figures, but 0.002 has just one (although it has three decimal places).

Mathematical processes in regard to decimal places and significant figures are straightforward. When you multiply two numbers, your result should always have the same number of significant figures as the smallest number of significant figures contained within your multipliers. So, for example, 2.11 × 3.13 × 10. is equal to 66.043. However, our least significant figure is 10.,[1] with two significant figures, so our result should be written as 66. If the first digit dropped is five or greater, round up; otherwise, round down.

When you add or subtract figures, your result should have the same number of decimal places as the least accurate figure has. Hence while 2.11 + 3.11 + 10. equals 15.22, since 10. is accurate to only two decimal places, our result should be written 15.00. As with multiplication, round the highest-value dropped figure up if it is five or more, down if it is less.

---

[1]Note that 10. has two significant figures because of the decimal point, whilst 10 has one significant figure, and 10.0 has three.

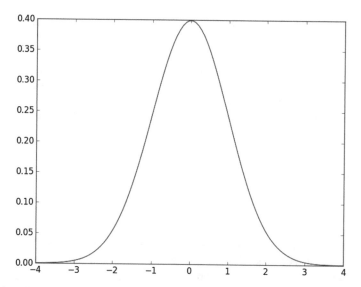

Fig. 11.1 A Gaussian plot. The centre of the peak is located at the mean value of the sample. The location of the peak is partly representative of systemic errors, while the width is characterised by the standard deviation of the sample

11.3 Handling Errors

One of the most common methods of finding uncertainties is to make multiple measurements and take the mean or median as the result and the standard deviation as the error. The standard deviation is also known as the root-mean-square deviation (RMS) for reasons that will become clear.

We can assume that in most cases, our random errors will have a Gaussian distribution, whose plot looks like Fig. 11.1. In this case, we can use the RMS to measure the distribution of the errors. The standard deviation of a sample is found by applying (11.1), where $\bar{x}$ is the mean of the sample, x_i is the value of the ith member, and N is the number of items in the sample:

$$\sigma = \sqrt{\frac{1}{N} \sum_{i=1}^{N} (x_i - \bar{x})^2}. \tag{11.1}$$

If our distribution of values is Gaussian, then 66% of our sample will be contained within one standard deviation (1σ) of the mean. At 3σ, it is 99.7%, and at 5σ, it is at the level considered to be proof, since 99.9% of the sample is contained within the boundaries. The square of the standard deviation is known as the variance and is a measure of the spread and is always nonnegative.

When you are looking at your data, beware of outliers. Outliers are results that appear on unexpected parts of the distribution curve, the plot of value against the count. Very large outliers can dramatically change your results, especially when you are dealing with small sample numbers. If you have an unexpectedly small or large result compared with the rest of the sample, you may wish to investigate why this might be. Often it is an indication of an actual error (i.e., a mistake has been made) rather than a true result. If you cannot find the mistake, you may exclude that result, but you must note it in your results, as it might have been significant. In such cases it may also be worth increasing your sample size to ensure that it is indeed a statistical anomaly.

11.3.1 Systemic Errors

As stated above, there are two kinds of errors: random errors and systemic errors. Systemic errors can also be divided into errors that can be addressed directly and those that cannot. For example, the bias added during the read process is largely unknown directly but is removed by the subtraction of a bias frame. Likewise, the dark current is an offset that can be largely subtracted by the dark frame, although there is an additional random element in the dark current. The nonuniformity of the illumination of the CCD and the pixel sensitivity is largely constant over short time scales and can also be considered in part systemic. Unfortunately, perfect flat fields are extremely difficult to produce and quickly become dated as dust collects on the optics. So although flat field subtraction deals with systemic errors, there is always a degree of random error inside a flat, which, as we will see, can be significant. However, systemic errors will always be of the same sign and magnitude for all your readings.

There may well be other unadjusted or unknown systemic errors. For example, the shutter on your camera might be sticking slightly so that your 1 s exposure is, in fact, 1.01 s. This will increase the number of photons you detect in a 1 s exposure. Likewise, you might be using a reference star to determine the magnitude of a star with an unknown magnitude. If the magnitude of the reference star is overreported by two magnitudes, you will have a two-magnitude systemic error.

Systemic errors are extremely difficult to detect. To avoid them, you should ensure that your instrument is correctly calibrated. If you suspect that you might have a significant systemic error, then compare your results with a known set. If your mean results constantly appear outside of the uncertainty range of the known results, you might well have an unknown systemic error, especially if they constantly appear on the same side of the known result.

A tale of caution. In 2011, the OPERA experiment mistakenly announced that they had observed neutrinos travelling faster than light. It turned out that an ill-fitting fibre optic cable was causing a systemic error, and in fact, the results were entirely consistent with previous experiments.

11.3.2 *Random Errors*

Random errors (also known as noise) are an inherent and unavoidable aspect of any system. However, you should endeavour where possible to minimise random errors by, for example, cooling the CCD, and quantify those you cannot.

In order to reduce errors, you should perform many observations if possible. As you can see from (11.1), increasing the sample reduces the standard deviation by $\sqrt{\frac{1}{N}}$. A simple way of determining whether your errors are large is to look at the spread of your sample, i.e., the maximum value minus the minimum value minus the mean value, and multiply this by 0.66. This will give you an approximate indication of your 1σ error, although you should beware of large outliers and small sample sizes, which tend not to be Gaussian in distribution.

A common trick used by marketers of beauty products is to say, for example, that 85% of women agree that the product has a positive effect. This result seems to suggest that the product works, when in fact, the sample size is likely too small for the results to be meaningful.

Many of your errors will take the form of **shot noise**, which is noise that is a result of the quantised nature of both electrons and photons. Shot noise has a Poisson distribution rather than a Gaussian. The Poisson distribution is characterised by the following equation:

$$P_\mu(\upsilon) = e^\mu \frac{\mu^\upsilon}{\upsilon!}. \tag{11.2}$$

In terms of astrometry, your random errors are largely caused by the pixel size and the seeing, which is the Gaussian distribution of the starlight caused by random changes in the atmosphere. There will also be errors from the distortion of the image due to the optics and problems inherent in the processes used.

You will often hear the term **signal-to-noise ratio**, or SNR, in astronomy. The SNR is just the ratio of the part of your signal (or count in the case of optical astronomy) that is coming from your image to that part that is coming from random fluctuation in the system. If you have low signal to noise, your uncertainties will be high, and vice versa. Generally speaking, an SNR of 100 is good, and 200 is excellent, but it depends on what you are trying to achieve.

Many of the astronomical applications used for photometry, and some for imaging, will report the signal-to-noise ratio (or just the noise of the background). You should use these figures with the understanding that they are underestimating the uncertainty. The measurement report is based on the variation in the sky signal. However, many things contribute to errors in observational astronomy, and although the sky signal is the most dominant source of error, it is not the only one. Flat field, dark, and bias noise, interpolation errors, and charge transfer errors as well as scintillation noise all contribute to the overall noise level.

When you are dealing with spectrography rather than imaging, the problem of noise becomes more important. Although we vertically collapse the spectra so that we have a single line of pixels with each pixel representing a small wavelength range, that pixel includes light from the source and sky lines—which are not uniform. The subtraction of the sky during the reduction process should remove most of this noise, leaving just the shot noise. Binning can help to improve the signal-to-noise ratio. By joining adjacent pixels, we increase the signal while, we hope, averaging out the noise to some degree, as the error scales with $\frac{1}{\sqrt{N}}$. However, as the signal intensity varies across the spectrum, so does the signal-to-noise ratio. Therefore, the signal-to-noise ratio of the spectrum is the mean pixel signal-to-noise ratio.

11.3.3   Working Out the Error Value

For photometry, the simplest method for determining the error is to use the reported signal-to-noise level (S/N). The S/N then becomes an error in magnitude by applying (11.3) below, where $\sigma(m)$ is the magnitude error, S is the signal in flux or counts, and N is the noise, also in flux or counts:

$$\sigma(m) = \pm 2.5 \log\left(1 + \frac{S}{N}\right) \qquad (11.3)$$

Equation 11.3 does not deal with errors that do not in some way link to the background variation. This is best addressed by taking repeat observations (assuming that the source is not intrinsically variable over short time scales). Once you have your measurements, find the range and then apply (11.4), where Δx is the error in the sample, x_{max} and x_{min} are the maximum and minimum values in the sample, and N is the sample size:

$$\Delta x = \frac{x_{max} - x_{min}}{2\sqrt{N}}. \qquad (11.4)$$

Equation 11.4 is not suitable for large samples, and we have to find the mean value of the sample and its standard deviation as in (11.1) and then apply (11.5), where Δx is the error in the mean of the sample, σ is the sample standard deviation, and N the sample size:

$$\Delta x = \frac{\sigma}{\sqrt{N}}. \qquad (11.5)$$

11.3.4   Propagation of Uncertainties

It may come as a surprise, but when you add two results, you do not add the errors. There are very specific mathematical rules that govern how to treat errors when performing mathematical operations on results. This methodology is known as the propagation of uncertainties.

When adding or subtracting two values x and y and their errors Δx and Δy to find z, the error in the resultant value Δz is given by

$$\Delta z = \sqrt{(\Delta x)^2 + (\Delta y)^2}.$$ (11.6)

For the multiplication of two values, the propagated uncertainty is

$$\Delta z = |xy| \sqrt{\left(\frac{\Delta x}{x}\right)^2 + \left(\frac{\Delta y}{y}\right)^2}.$$ (11.7)

For the division of two values, the propagated uncertainty is

$$\Delta z = \left|\frac{x}{y}\right| \sqrt{\left(\frac{\Delta x}{x}\right)^2 + \left(\frac{\Delta y}{y}\right)^2}.$$ (11.8)

When raising a result x to a power n, the error in the result z is given by

$$\Delta z = |n| x^{n-1} \Delta x.$$ (11.9)

When applying a function f to x and y, the propagated error is given by

$$\Delta z = \sqrt{\left(\frac{\delta f}{\delta x}\right) + \left(\frac{\delta f}{\delta y}\right)(\Delta y)^2}.$$ (11.10)

Chapter 12
High-Precision Differential Photometry

Abstract It is possible, but challenging, to achieve the millimagnitude uncertainties needed for research-level time series photometry with small telescopes. In order to do so, the telescope–instrument uncertainties need to be evaluated and addressed, the check stars must be carefully selected, and nonstandard methods such as defocused photometry should be considered.

12.1 Introduction

In recent years, with improving noise levels and the development of high-sensitivity low-cost CCD, and now CMOS, astronomical cameras, the new science of exoplanet detection has given small telescopes, such as those found in teaching observatories, a greatly extended utility. Further, the capacity of small telescopes to perform milli-magnitude photometry, in far from ideal environments, has enabled their repurposing to a new research life.

This chapter is largely derived from excellent observational work undertaken by the late Peter Beck at the University of Hertfordshire. If you wish to read more details, his thesis can be found online. The chapter focuses on achieving millimagni-tude differential photometry in the context of exoplanet transit observations through the improved understanding that comes through the calibration of the telescope and instrument package. However, there are lessons here for anyone wishing to under-take astronomical photometry. Many of these processes should be undertaken by observatory staff if your observatory is staffed.

There are four fundamental sources of noise in photometric measurements:

- Poisson noise in the target data and the check stars.
- Calibration noise within the bias, dark, and flat-field frames.
- Scintillation noise caused by the effects of the atmosphere.
- Read noise caused by the electronics and cables, with electromagnetic interference in the observatory.

© Springer Nature Switzerland AG 2020
M. Gallaway, *An Introduction to Observational Astrophysics*,
Undergraduate Lecture Notes in Physics,
https://doi.org/10.1007/978-3-030-43551-6_12

All four of these factors contribute to the overall uncertainties in your photometric observations. Additionally, there are other uncertainties such as the variability in the check stars and atmospheric conditions such as high cirrus.

12.2 Calibration Noise

One of the most controllable forms of noise is in the calibration of the images. The most obvious way to start is to ensure that any bias, dark, and flat-field frames you are using are as up to date as possible. Bias frames are the most stable over time. Dark frames are relatively stable, but they need to be taken regularly and whenever there is an equipment change in the observatory that might cause additional noise, for example any cable change or a power supply change. Flat-fields should be done every day if possible. Historically, I have found that super dome flats produced using a flat-field luminescent panel tend to be more stable than others. You should always use super bias, flat, and dark calibration frames when processing science frames for high-precision photometry. These are produced by taking the mean of several (at least ten) frames. Recall that dark and flat field frames are temperature specific and that flat frames are also filtered variant. Also, be aware that on a warm day, camera coolers work far harder than in the evening and can produce some additional read noise, so it is best to take dark and flat fields after sunset.

Dark current is proportional to both sensor temperature and integrated exposure time. You should try to cool the sensor to the lowest temperature that your imager cooling system can comfortably cope with. It is common to use scaled darks for imaging. This means that if you don't have a 60 s dark for your exposure but have a 30 s dark, you subtract the 30 s dark from your image twice. This works, and for imaging it is a perfectly acceptable practice. It is also acceptable for photometry, but when you are trying to reduce your uncertainties to the limit, it is best to have the correct dark frame.

The total noise contribution from the calibration frames is the square root of the sum of the squares of the standard deviations of the bias, dark, and flat-field frames.

12.3 Bias and Dark Noise

Dark and bias frames are used to remove noise from the read process and thermal electrons from a science frame. However, the dark noise is effectively random, as is to some extent the bias noise. Neither calibration frame can remove all the noise, and in fact, in a low noise calibrated frame, calibration frames might introduce noise. Hence, the noise contribution by dark and bias frames is not the standard deviation with the calibration frame but the standard deviation pixel to pixel between multiple calibration frames.

To obtain this figure, a large number of calibration frames need to be generated and compiled into a three-dimensional FITS cube (FITS files can be multidimensional) with axial directions x and y being the horizontal direction and vertical direction of the individual frames, and axis z being the individual calibration frames. By taking the standard deviation through the z-axis, you obtain a two-dimensional collapsed FITS frame. There are several methods you may use to determine the noise within the calibration frame set. The author normally takes the mean value of the collapsed frame as the noise value for the set.

To find the noise within the bias and dark calibration frames, we cannot just find the standard deviation between our frames, as these include hot, cold, dead, and stuck pixels. The variability of pixel sensitivity is addressed by the application of a flat field frame and will be part of the flat field noise. As hot, cold, dead, and stuck pixels always give the same value, they do not generate significant noise in time series differential photometry, as we are looking at the change in brightness, and these pixels do not change in value. This comes, however, with the proviso that the target and check stars have not drifted in the image, a problem that is discussed later in the chapter.

Typically, a dark or bias frame should have a low mean pixel count, as we are detecting only read noise and thermal electrons. This means that pixels stuck at a low value (often zero) or high value (often the saturation limit) will report a cross frame standard deviation of zero. This, in turn, will falsely reduce the calibration noise figure we are trying to find. We therefore have to remove the bad pixels before calculating our final uncertainty.

There are two points in the process at which we can do this. Either we can address the problem at the individual frame level or we can address it at the final, collapsed, frame level. Computationally, the second method is easiest, and we have to perform the processes only once. In both cases, you will need to perform some kind of clipping, whereby values outside of a given range are discarded or replaced with another value, normally the mean. This might be **sigma clipping**, whereby data with values outside a specified number of multiples of the standard deviation are moved or replaced. Sigma clipping can be dual-tailed, where the top and bottom ranges are removed, or single-tailed, where only one end is removed. An alternative method is a numerical clip, where values outside a specified range are clipped. This range is normally determined from a frequency plot of the pixel values and, like sigma clipping, can be at either end or both ends of the pixel value range.

Clipping the collapsed frame has the advantage of being faster, but pixels that are inherently low noise might be excluded, as they will have a low standard deviation (as will most bad pixels, although for different reasons).

Another alternative is to identify the bad pixels at the single-frame level using a reverse sigma clip, whereby only bad pixels are left, and then use this as a mask to apply to the collapsed frame.

Whichever method you choose, there is, at the time of writing, no single application that can perform all the steps needed. However, it is relatively straightforward to produce a Python script using AstroPy and Numpy to achieve this process, and it is a good project as an introduction to astronomical programming and FITS manipulation.

The combined contribution from dark and bias noise is likely to be insignificant, and one would expect it to be significantly outweighed by noise from the flat-field frames and the sky. If, however, you do have a high-level noise in the dark and biases, there are several things that can be done to address them.

Firstly, check that your analysis of the errors is correct, for example by performing the same operation on files from a good instrument. For example, data can be downloaded from the ESO website. If the uncertainty is real, then it is best to dismount the instrument from the camera and test it in a lab, away from any electrical noise source. If possible, try replacing the camera power supply and data cable with a known low-noise version. If in this setup the noise becomes low, it is most likely a faulty power supply or cable. If the noise persists, then it is likely an inherent fault with the camera, which might not be easy to resolve.

12.3.1 Flat-Field Noise

Flat-field noise cannot be directly measured, as the noise is the variation in the flatness of the image frame after the application of the flat field.

The noise in a calibrated frame is given by

$$\sigma_{cal} = \sqrt{\sigma_{Bias}^2 + \sigma_{Dark}^2 \sigma_{Flat}^2}, \tag{12.1}$$

which we can rearrange as

$$\sigma_{Flat} = \sqrt{\sigma_{cal}^2 - \sigma_{Dark}^2 - \sigma_{Bias}^2}. \tag{12.2}$$

Hence, we can get to the flat-field noise by measuring the noise in a calibrated frame. Ideally, you need a calibrated frame that has few stars in the image. The noise within the frame will have to be measured away from any star or other astronomical objects in the field, although this is not a problem for dome flats. Again, this may be achieved by sigma clipping the calibrated frame, or the region function of SAO DS9 can be used to select and analyse a region of the calibrated frame free from stars. By applying (12.2), we obtain the flat field noise σ_{Flat}. You may wish to use multiple images and multiple flat-field frames to obtain a more accurate uncertainty reading.

Poorly implemented flats-field frames can be a significant noise contributor. Good flats take a considerable time to undertake, and if twilight flats are being used, the area of the sky selected is often suboptimal and may vary in brightness. Although sky flats are normally considered the best method to obtain flats, improvements in flat-field lighting sources have closed the gap between sky flats and dome flats.

A typical dome flat is taken using the inside of the dome. This is unlikely to be a uniformly reflective surface, and the dome lights are unlikely to have a flat spectrum.

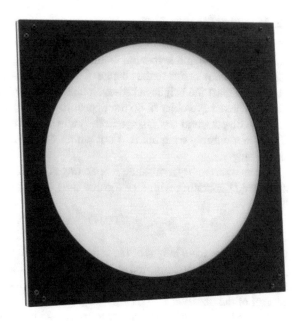

Fig. 12.1 An example of a flat field box, designed to give a very uniform and spectrally flat light source for making flat fields. (Image courtesy of Artesky Italy)

Other methods such as placing a largely white matte board on the dome for use in dome flats may be an inexpensive method of improving flat fields; as might covering the telescope aperture with white semitransparent fabric (I've seen old t-shirts used, but I would not recommend this). However, currently, the best method for dome flats is to use a specifically designed electroluminescent board known as a flat field box. These are designed to fit over the telescope's aperture, completely obscuring stray light and providing a uniform light source. They also tend to be bright, so that flat fields are completed faster, so that the super flat field can be drawn from more frames. It also has the advantage that if somebody walks into the dome while the flat fielding is taking place, the exposure in not affected, unlike sky flats or dome flats, which use the dome lights. You will be surprised at the number of academic staff who ignore warning signs not to enter a telescope dome while it is flat fielding.

As stated previously, the ability of flat fields to adjust for imperfections in the optics of the telescope and the sensitivity of the imager is time-sensitive, much more than for dark and bias frames. If the camera is removed and refitted or repositioned or a filter changed, the flat fields must be redone. Ideally, flats fields should be done every day there is observing, although in reality, this is hardly ever done, especially in a small teaching observatory. However, if you wish to undertake very low noise photometry, regular and frequent flat fielding is a requirement (Fig. 12.1).

12.4   Scintillation Noise

Scintillation is what a layperson would call twinkling, a change in brightness of an astronomical object due to the change of pathway caused by atmospheric turbulence. Planets, at least the bright naked-eye planets (not counting Uranus, which is technically a naked eye object), do not appear to twinkle, because unlike stars, they are resolved objects, and any change in the pathway is limited to the disk of the planet. This is not entirely true, but it is not within the ability of the human eye to detect such changes.

For an observatory at altitude h operating at an air mass X, with an instrument of diameter D operating with an exposure time of t_{exp}, the scintillation noise is given by

$$\sigma_{scint} = 0.004 D^{-\frac{2}{3}} X^{\frac{7}{4}} e^{-\frac{h}{H}} (2t_{exp})^{-0.5}, \tag{12.3}$$

where H is the scale height of the turbulence, which can be taken as 8000 m.

Equation 12.3 is a simplified and generalised scintillation noise equation. It ignores such effects as wind speed and turbulence cell size, as these are generally out of the control of the observer.

As can be seen, the best practice to reduce scintillation noise is always to observe the target as it crosses the meridian and has the lowest air mass. However, for observations such as exoplanet transits this is impossible, as we are constrained to the period of the transit, and long-period observations are going to be over a range of air masses, with the highest scintillation noise occurring at the highest air mass. Increasing the diameter of the telescope would appear to be the solution to this problem. However, increasing the diameter of the telescope reduces the exposure time by the square of the diameter. As we can see, increasing the exposure time decreases scintillation noise, but we can expose only up to the point of linearity of the instrument, t_{max}. The solution to this problem is to bin the data postcalibration, sacrificing cadence for reduced scintillation noise.

Increasing the aperture size of the telescope has an additional effect. The turbulence cell size is typically of order 30–50 cm. Hence, telescopes significantly larger than the cell size have less scintillation noise, as it is more likely that the entire range of possible photon pathways will reach the camera.

Returning to (12.4), we can see by increasing the exposure time, we reduce scintillation noise (although by the square root of the exposure time), although of course, we would not be able to exceed t_{max}.

A possible solution to scintillation noise is to implement a method known as **defocused photometry**. The telescope is deliberately defocused to allow an increased exposure time. This method works because defocusing spreads the photons over a larger area, and hence the light falls on more pixels. Therefore, a longer exposure can be made while keeping the CCD in its region of linearity. Defocusing can reduce, but not eliminate, scintillation and other atmospheric effects as well as flat-field errors. You should be aware that defocusing can increase sky noise, so care must be taken to keep the target's PSF only as large as required. It is also impossible to automatically plate solve a defocused image, as the plate-solving algorithm will not be able to identify the stars in the field, as they are no longer Gaussian. Likewise, any automatic photometry is also going to be impossible for the same reason. It is also likely that you will not be able to defocus during remote or robotic operations.

Defocused photometry is more susceptible to bad and poor pixels (as the region of the CCD/CMOS is larger) and larger dark noise due to the extended exposure time, although this may be offset by lower scintillation noise. It also is not suitable for remote and robotic observations, as they require the telescope to remain in focus.

12.5 Sky Noise

Sky noise is caused by changes in the emissivity of the atmosphere. Although the atmosphere is largely transparent in the optical range it still emits a small amount of light as well as reflecting light from the ground and astronomical objects from dust particles in the atmosphere. In a well calibrated image, devoid of bad pixels, with zero contribution from bias, dark, and flat-field sources, the count away from a source represents the sky signal. You should notice when observing when the Moon is up or when imaging close to a bright planet that your sky signal is higher.

As explained in Chap. 10, when performing aperture photometry, we sample the sky signal using an annulus around the target (or away from the target if there is another object in the annulus) and use the mean value of the sky in the annulus as the sky value, which in turn is subtracted from the source count. The variation in the sky signal is the **sky noise**. Sky noise is the main limiting factor in observing faint objects. Sky noise is broadly equal to the root sky count, as sky noise is generally Gaussian, especially for longer exposures. Hence, increasing the exposure time does not improve the impact of sky noise on the uncertainties in your photometry. If your target signal does not exceed the sky noise by at least a factor of two, you are unlikely to achieve reliable detection.

When you perform aperture photometry using an application such as APT or Maxim DL, the reported uncertainty produced by the software is based on the sky count. The software assumes that noise from other sources has been addressed, although in fact, in most cases it has not. Effectively, the software is giving you a "best case" uncertainty. There are a few simple methods that can be undertaken to reduce sky noise. The most obvious one is to site the observatory somewhere dark and away from local population centres. This is not that practical for many teaching observatories where the campus is in a big city. Almost all London-based universities that teach astrophysics have sky noise issues from city lights as the city has grown around them. Sky noise also scales with altitude, which is why many of the world's best observatories are at high locations, and the Hubble Space Telescope can perform so well despite having only a 2.4 m mirror.

Observing your target at low air mass and when there is no Moon or when the Moon is at a crescent phase will also improve sky noise, as will avoiding bright stars and planets. High cirrus clouds, invisible from the ground, also increase sky brightness.

Some observatories have sky temperature meters that measure the temperature of the sky by looking at its infrared brightness. A warm sky is a key indicator of high cirrus clouds (although not always if they are very high). Likewise, many observatories have a sky brightness meter, which is effectively a small telescope with a camera pointing at the meridian and measuring the sky count for precision photometry, you should try observing only on nights with low sky brightness.

However, as noted previously, phenomena such as exoplanet transits tend not to be at optimal times where there are good photometric sky conditions. Given that you are unlikely to be able to move your telescope away from city lights or to a high altitude, the only real way of dealing with sky noise is to choose the right photometric filter. Sky noise is relatively flat across the optical part of the spectrum, except for some strong lines from sodium and other street lighting sources, but your target is going to be brighter in some bands than others (which can be checked in SIMBAD). Imaging in the brightest band will improve the signal to sky noise ratio. So for example, if your target is an M-dwarf, you should image in the R band (or a similar red end band).

12.6 Lucky Imaging and Adaptive Optics Photometry

Lucky imaging was introduced in Sect. 8.8.1 as a process whereby very high angular resolution imaging may be accomplished using a modified commercial webcam or an astronomical webcam. These standard webcams are unsuitable for photometry, as they have high noise levels, especially read noise. However, a new generation of lucky imagers based on **electron multiplying CCDs**, or EMCCD, are now being deployed. More often used on high-magnification shallow depth of field microscopes, EMCCD use a photomultiplier to increase the signal without increasing read noise.

By examining the noise within the image and the point spread function of the stars therein, it is possible to reduce the images taken to those with low scintillation and sky noise. Stacking these to form single images with a cadence similar to that produced by a traditional imager should produce images with lower noise and hence lower uncertainty.

EMCCDs are heavy compared to traditional imagers and therefore can be difficult to mount on a small telescope. Their very high frame rate (the author has used an EMCCD at a frame rate in excess of 1000 frames per second) means that standard computer interfaces are too slow to cache the data, and special interface cards are needed. Also, the rate of caching can exceed the transfer rate to the computer's hard drive.

An additional problem with EMCCDs is that they can be damaged when pointed at bright sources. This includes the Moon, most planets, and bright stars.

Lucky imaging photometry is out of the range of small observatories and is still in its infancy. The signal-to-noise gain to cost is high, although for crowded field and PSF photometry, lucky imaging may provide advantages over standard imaging systems.

Adaptive optics (AO) is a process involving the variation of the optics of the imaging device in response to changes in the focusing of the optical system caused by, in astronomy, atmospheric turbulence. To do this, a source in the field of view needs to be monitored, and an active feedback system makes corrections to the optical system, either a flexible mirror or lens in the optical pathway, in order to make the point spread function of the observed target as small as possible. In low-cost systems that are now being made available to amateurs and small observatories, the source will be a star. At large observatories such as the VLT in Chile, a laser tuned to make a visible spot at a specific layer in the atmosphere is used.

As the point spread function is reduced, we obtain the effects of improved seeing and finer resolution of the telescope. This results in the target being spread over fewer pixels, resulting in a reduced exposure time before we enter the nonlinear zone of the detector. As we have seen, increasing the exposure time will decrease the uncertainness. As a result, AO is not suitable for aperture photometry in uncrowded fields. In crowded fields, where aperture photometry is challenging, AO may reduce the crowding to the point where aperture photometry is possible. However, in crowded fields, PSF photometry is the preferred photometric technique, and AO should increase the reliability of PSF photometry in these fields.

12.7 Target and Check Star Noise

As light comes in discrete packets, photons, in any given period there is a chance of receiving more or fewer photons than the mean number from a source of constant luminosity. This is known as **shot noise**, and it takes the form of a Poisson distribution, whereby the noise increases with the square root of the signal. Hence, the signal-to-noise ratio for source and check stars scales by

$$SNR = \frac{N}{\sqrt{N}}. \tag{12.4}$$

Hence the only way to reduce target noise is to increase the signal by increasing exposure time.

Check star noise, however, can be addressed in an alternative manner. Adding more check stars decreases the overall check star shot noise by $\frac{1}{\sqrt{N}}$. But beware! Apart from the fact that there are diminishing returns (a sample of four will have half the error of a single sample, but a sample of nine, over twice the work, will reduce the uncertainty only by a third), there is an increased risk of a bad pixel or, perhaps worse, a variation in the brightness of a check star.

This last point, magnitude variation in the check stars, is resolvable by careful selection. We can pick a section of stars in the field with magnitudes similar to the target's magnitude and perform photometry on them. For each image within your time series, find the mean count of the check stars and plot them. The plots should be smooth, with a slight gradient caused by changing air mass. If there are spikes within the data, that is an indication of a problem with one or more of the check stars. Even if there is no indication of a problem, it is good practice to plot each of the check stars along with the mean. If they are significantly different from the mean, they should be removed from the sample and the mean recalculated. Likewise, if there is a change in brightness in one target star that is not reflected in the others (although it will be reflected in the mean), that should also be removed. What will remain after several iterations of this process is a set of check stars that are representative of the conditions being observed.

The photometric uncertainty is then

$$\Delta m = \sqrt{\sigma_{\text{ref}}^2 + \left(\frac{N}{\sqrt{N}}\right)^2}, \qquad (12.5)$$

where σ_{ref} is the standard deviation between the count in the check stars and the mean count of the check stars, and N is the count of the target.

12.8   Postmeasurement Uncertainty Reduction and Light Curve Folding

As we have seen in this chapter, there are many sources of noise in photometric systems, and by careful calibration and planning, you should have managed to reduce the uncertainties to the point that the signal-to-noise ratio is great enough for an observation to be successful. If that is not the case, there is a single course open to you: to sacrifice cadence in order to improve photometric accuracy. However, be aware that by doing so, you will increase the time-series uncertainties, which are the times between the start of one frame and the start of another (not the exposure time), which might result in some important features in the light curve being lost. By binning adjacent data points, taking the mean of the two points, the uncertainty is reduced by 70%.

If you are observing a regularly occurring feature such as an eclipsing variable or an exoplanet transit, an alternative method to improve photometric accuracy is **light curve folding**. Light curve folding works by overlaying repeated observations on a single curve. For this to happen, you need to identify a feature that occurs in all the light curves, which typically will be a contact shoulder but not always. Note the chronological position of this feature in each of the light curves; the FITS header should have this as a Julian date. If you know the period between events, you can fold the light curve into **phase space**. This can be done in a programming language like Python or simply in a spreadsheet. The data point with the lowest Julian date (i.e., the first one taken) is used as the anchor point in this illustration. It can be changed if you wish. We shall call the Julian date of this observation t_0. We can now determine the phase space of all other data points using

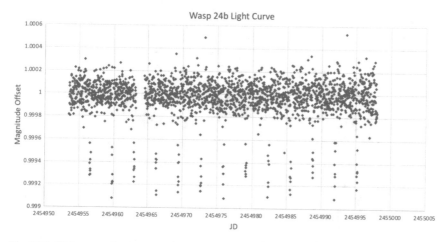

Fig. 12.2 Light curve for exoplanet Wasp-24b plotted using data from the SuperWASP archive

$$\rho_n = \frac{t_n - t_0}{P} - \text{int}\left(\frac{t_n - t_0}{P}\right), \tag{12.6}$$

where ρ_n is the phase space location for the nth data point, which has a Julian date of t_n, and the period is P. Plotting the phase space against the change in flux or magnitude gives the folded light curve, like the one seen in Fig. 12.3.

Figure 12.2 is a light curve generated from SuperWASP photometric data. There are 2100 data points with a minimum cadence of 1765 s. The period of the plot covers 44 days. We have removed the errors to avoid confusion due to a large number of data points. As can be seen, there are signs of multiple transits. However, the high cadence period makes the identification of any structure within the light curve difficult to detect, apart from the transit itself and the period between transits.

Figure 12.3 shows the same data as Fig. 12.2, but in this case, the data have been folded, using the established period between transits. Rather than representing time, the horizontal axis now represents the phase angle. We now see the transit more clearly and can see features, such as first and last contact, that can be used to determine the Wasp-26b's diameter. We have also offset the phase somewhat from the first data point in order to put the transit close to 0.2 in phase space for easier reading of the plot (using the first data point in the set would have put the transit at the end of the plot). This demonstrates the importance of multiple observations and the power of light curve folding. It also shows why high cadence is desirable. The long period between observations results in a significant loss of detail in the individual transits, and we get significant detail only once we fold data containing 13 transits over 44 days. It should be remembered that the SuperWASP telescopes are exceptionally small and cover a very wide field of view, and consequently it is difficult to avoid long cadence periods. They are also highly successful survey instruments that have been eclipsed only by Kepler.

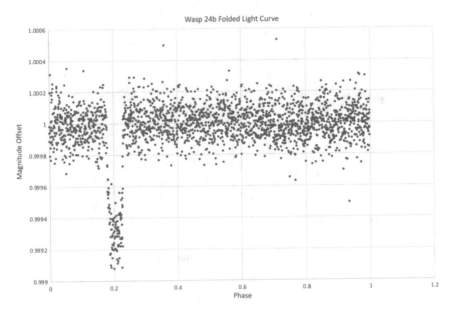

Fig. 12.3 Folded light curve for exoplanet Wasp-24b plotted using data from the SuperWASP archive

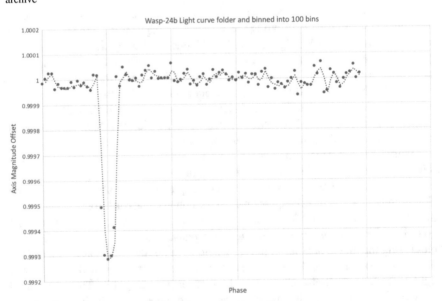

Fig. 12.4 Folded and binned light curve for exoplanet Wasp-24b plotted using data from the SuperWASP archive

Figure 12.4 shows the result of Fig. 12.3 being resampled into 100 bins. As can been seen, the scatter in the magnitude change has improved, indicating that the uncertainties have been reduced, allowing a reasonable trend line to fit. However, the reduced number of data points within the transit second of the plot means that we have lost some details that are of some interest. This example shows that reducing the uncertainties by binning has a significant effect on the cadence, especially in this case, where the cadence is in excess of 1700 s. Increasing the number of bins does not in this case significantly improve the details in the light curve.

The light curves above illustrate that when you are folding multiple observations into a single phase space, a plot is a powerful tool in time series observations. It also illustrates that although binning data is a viable approach to decreasing uncertainties, it is not suitable for all observations, especially with time-series observations with low cadence.

Chapter 13
Solar Astronomy

Abstract The Sun is the most obvious astronomical object and the nearest star. It is dynamically active, with many changing surface features. We discuss sunspots, flares, prominences, and other solar features and, via a practical example, show how to view the Sun in a safe manner. We conclude with a brief discussion of the nature and observation of solar eclipses.

13.1 Introduction

With a diameter 109 times that of Earth and containing 99.89% of the mass of the solar system, the Sun dominates the solar system and our daily lives. It provides almost all the energy for biological processes on Earth, and its long-term, and arguably short-term, evolution has a major impact on life and human civilisation.

The Sun's composition is typical of a star of its age, 75% hydrogen, 24% helium, and a smattering of other elements, including carbon and oxygen. High densities and temperatures within the Sun's core ($150 \, \mathrm{kg \, m^{-3}}$ and $15 \times 10^{15} \, \mathrm{K}$) create the environment for a sustained thermonuclear reaction, the conversion of hydrogen into helium via the proton–proton chain. The first part of this chain, the creation of deuterium, is the choke point in the process, and it is sensitive to conditions. When the core is exceptionally hot or dense, more deuterium is created and more energy released. The release of more energy causes a slight expansion in the core, dropping the core temperature and pressure and thereby suppressing the fusion rate. Hence we have a feedback loop whereby the radiation and thermal pressure within the Sun are matched by the gravitational pressure; the Sun is in hydrostatic equilibrium.

The nuclear synthesis of helium from hydrogen results in the liberation of gamma rays and neutrinos. Neutrinos weakly interact with matter and escape the Sun in a few seconds. The gamma rays, however, are continually absorbed and reemitted as they pass through the layer above the core—the radiative zone—and are converted into lower-energy photons. Eventually, the radiative zone gives way to the convective zone, and energy is now transported to the surface by the convection of hot gas. Once the photosphere, the visible surface of the Sun, is reached, the gas temperature has

M. Gallaway, *An Introduction to Observational Astrophysics*,
Undergraduate Lecture Notes in Physics,
https://doi.org/10.1007/978-3-030-43551-6_13

dropped to 5,500 K, and the energy created in the core is reemitted as optical and ultraviolet light. The typical time taken for a photon to escape the Sun via this process is of order a million years.

The atmosphere of the Sun above the photosphere consists of several layers hosting many of the events you will observe. The coolest layer is that immediately above the photosphere. It is at about 4,000 K, which is cool enough for some molecules to exist. It is this region that generates many of the molecular spectral lines we see in the solar spectrum (see Chap. 14). Above this cool layer is the hot region known as the chromosphere. This region is the source of many emission lines, including the important hydrogen alpha line. Above this lies the transition zone, a region of very rapid temperature rise, which is largely visible in the ultraviolet. The transition zone is, in turn, enveloped by the corona, an extremely hot region, up to 20 million K, which can be considered the extended atmosphere of the Sun. The corona is not normally visible to ground-based observations except during a solar eclipse.

13.2   Solar Features

In the practical section of this chapter, you will be asked to observe a number of solar features. These can be placed into two groups of features: photospheric and chromospheric. How you observe them depends on their location and their temperature.

13.2.1   Sunspots

The Sun has a very powerful magnetic field that drives many of the processes that we see on the photosphere. It is believed that the source of this field is the tachocline, the boundary between the radiative and convective zones. The radiative zone and core rotate as a solid body, while the convective zone and the boundaries above rotate differentially. This differential rotation drives a magnetic dynamo that is the source of the Sun's magnetic field.

Magnetic field lines escape the photosphere and loop high into the Sun's atmosphere before descending and reentering the photosphere. These field lines suppress convection, preventing the flow of hot plasma from below. The reduced convection results in a temperature drop at the photosphere of order 500 K. Although the photosphere is still hot, about 5,750 K, the 500 K temperature drop is enough for the region of suppressed convection to look dark against the bright photosphere. This dark spot is a sunspot. Sunspots always appear in pairs (although that is not always noticeable to an observer), reflecting the north and south poles of the magnetic field lines. The order in which the sunspots appear—for example, a north spot may be in front of a south spot in the direction of rotation—is reversed between solar hemispheres. The Sun goes through a period of increased magnetic activity, characterised by an increasing number of sunspots, every 11 years, a period known as the sunspot cycle. After

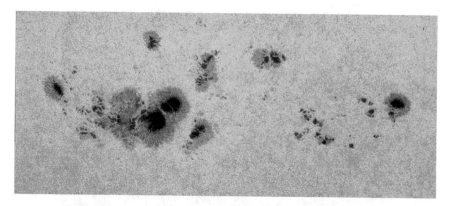

Fig. 13.1 Image of a large sunspot archipelago captured by Alan Friedman on 10 July 2012

a new cycle has begun, sunspots slowly reappear with their order reversed. Hence, there is a double cycle comprising two 11-year sunspot cycles. The reasons behind the 11-year and 22-year cycles are still unknown. However, if you are observing the Sun, it is useful to know where in the sunspot cycle we currently are.

Sunspots can be seen in a number of filters and are best seen in white light. Figure 13.1 shows a large group of sunspots observed in white light. The very dark area is known as the umbra, while the surrounding paler area is the penumbra. Spectrographic observations of sunspots show Zeeman line splitting caused by the intense magnetic field. The similarity of the region surrounding the sunspot to iron filings near a magnet should not be overlooked.

13.2.2 Faculae

Faculae are bright patches located just above the photosphere. They appear in areas of intensive magnetic activity, and it is this activity that is thought to cause their brightness, being 300 K above the surrounding photosphere. Hence a facula may be considered the reverse of a sunspot. Faculae are challenging to observe, and such observations are best undertaken near the solar limb. Figure 13.2 shows a faculae group.

13.2.3 Granules

Granules are small-scale structures found on the photosphere. They are convection cells of hot material rising from deeper within the Sun. They are dynamic structures,

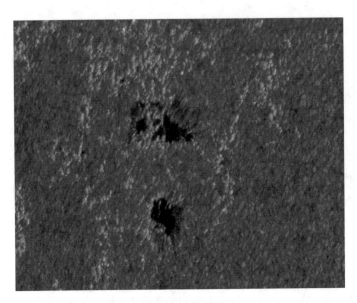

Fig. 13.2 Image of faculae (*white*) on the surface of the Sun surrounding a sunspot pair. (Image courtesy of NASA)

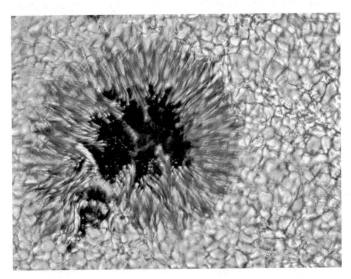

Fig. 13.3 Image of convection granules surrounding a sunspot. (Image courtesy of NASA, Robert Nemiroff (MTU) & Jerry Bonnell (USRA))

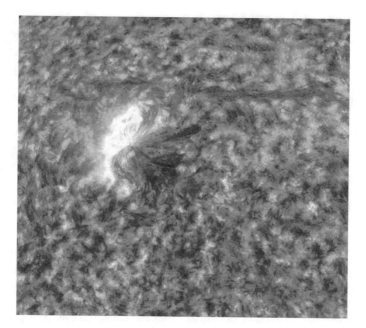

Fig. 13.4 Image of solar plage. (*Image credit* NASA/SDO)

changing rapidly in comparison to other solar features. The dark edges are the cooler gas descending. Figure 13.3 illustrates a typical granulated solar surface.

13.2.4   Plage

Plages are bright extended beach-like features (hence the name) located above and around sunspots. They are very bright in hydrogen-alpha, which leads to the observation that although in white light sunspots darken the Sun, in hydrogen-alpha, the presence of plages results in brightening. Figure 13.4 shows a plage sitting over, and obscuring, a sunspot pair.

13.2.5   Filament

13.2.6   Prominences, Filaments, and Flares

Given that the surface of the Sun is a plasma, it is not surprising that there is an interaction between the magnetic field lines extending from the sunspots and the hot

Fig. 13.5 Image of solar prominences. (*Image credit* NASA/SDO)

plasma. The plasma becomes trapped in the magnetic loop and enables the observer to view the loop extending high into the chromosphere. When viewed from the side, such a loop of trapped gas is known as a prominence (Fig. 13.5). When viewed from above, they block light from the hotter gas below, creating a dark line across the surface of the Sun. In such an orientation, they are known as filaments (Fig. 13.6). They are, however, the same feature. Stresses form in the field lines, caused by the loading of the trapped gas. Eventually, the stress on the field lines cause them to break and reconnect into the surface of the Sun, releasing a very large amount of energy in the process. The energy release is often accompanied by a sudden brightening in the trapped gas—a solar flare. Often, a large amount of coronal material may be ejected into space at high speed: a coronal mass ejection. If this material interacts with the Earth's magnetic field, it can lead to bright and extensive aurorae.

13.3 Observing the Sun

The observation of the Sun is one of the most hazardous activities an astronomer can undertake from the point of view of risk to both equipment and self. The Sun should never be observed with the naked eye, nor even with sunglasses.

The safest way to observe the Sun is by projection, with the use of a pinhole camera, a telescope, or a purpose-built instrument such as a sunspotter. These instruments focus the Sun's light onto a white surface, and it is the reflected light from that surface that is observed, not the Sun directly. When you align a solar projection instrument, it is best, as it is with all solar observing, to use the telescope's shadow to point at the Sun, rather than looking at the Sun itself. This method is safe, although it is difficult to record the images, and it shows only limited photospheric details, almost exclusively sunspots.

Solar filters are unlike normal astronomical filters. They are designed to cut out a significant amount of light over a broad range of wavelengths, including UV and

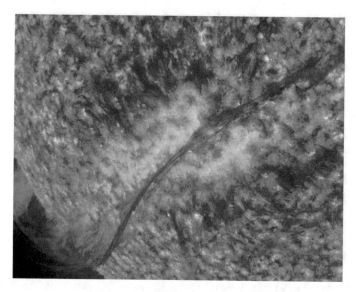

Fig. 13.6 Image of solar filament. (*Image credit* NASA/SDO)

IR. You should not use traditional, nonsolar filters except when using them behind solar rated filters.

Filters are either white light, i.e., they just reduce the amount of light that passes through to the observer across the spectrum, or they are narrow-line filters. If you suspect that a filter is damaged in any way, do not use it; even a pinhole could damage your eye or the camera. This is especially true of film-based filters such as those produced by Baader. Before using film-based filters, I always inspect them for holes using a bright light source (but not the Sun).

Hershel wedges are becoming increasingly popular for solar observing. The wedge consists of a prism that reflects most of the incoming light away from the observer and absorbs the unwanted UV and IR components. Hence, Hershel wedges are white light devices. However, they should always be used in conjunction with other light-reducing filters such as a polarizing filter or a neutral density filter, as Hershel wedges do not remove enough UV and IR for safe operation without postfiltering.

Specialist white light solar filters that mount over the primary aperture are available. These filter out the harmful light before it reaches any optical component. They tend to use an aluminised coating on glass and represent an advance over mylar film filters, as they are less prone to damage. When used with a webcam and a refractor, such a filter can be used to make high-resolution images of photospheric activities such as sunspots and granules.

For observations of the chromosphere, you will need either a specialist hydrogen-alpha telescope, such as manufactured by Lunt or Coronado, or hydrogen-alpha filters that can be mounted on a suitable refractor. In both cases, there will normally be filtering in front of the first optical component of the telescope and after the last.

Fig. 13.7 Lucky image of the Sun taken in hydrogen-alpha, showing sunspots and filaments. (Image by David Campbell, University of Hertfordshire)

Both of these will often have an adjustable degree of polarisation for the amount of light reaching the camera or the eye. Such filters are ideal for the observation of flares, prominences, plages, and sunspots.

An alternative to the hydrogen-alpha filter is the 393.4 nm calcium K line filter. This isolated light comes slightly lower in the chromosphere and has strong association with magnetic features, such as sunspots and faculae.

Given that you are observing the Sun, there is going to be considerable atmospheric turbulence, and often solar video files show shimmering or boiling, which reduces the resolution. By using a webcam, preferably an astronomical one such as built by DMK or Point Grey (there is no advantage in using a colour camera), one can use lucky imaging to obtain extremely high resolution images of the Sun, such as that shown in Fig. 13.7.

13.4   Practical 9: Exploring the Physics of the Sun

The dominance of magnetohydrodynamic forces in the observable layers of the Sun—the photosphere and the chromosphere—makes the Sun a dynamic and ever-changing target for astronomy. It is also the only star in which we can view these forces in any detail. Understanding the nature of surface features and their interactions is a key component of solar physics.

13.4.1 Aims

In this practical, you will be observing the Sun over a number of days to determine the evolution of photospheric and chromospheric features, understand limb darkening, and determine the Sun's differential rotation rate. In addition, you will learn how to observe the Sun in a safe manner.

13.4.2 Planning

You will need to observe the Sun over a period of approximately 14 days in order to determine the Sun's rotational period. You may have to overcome inherent safety features that prevent the telescope from pointing at the Sun. Before observing, you should ensure that the correct filters are fitted and that you understand the procedures for solar observing at the facility you will be using. It will be beneficial if you can observe in both hydrogen-alpha and white light if at all possible. The author suggests that you use a webcam as your image device and that you undertake lucky imaging in order to improve resolution.

13.4.3 Observing

Ensure that the solar telescope is correctly set up and the required filters are in place and are undamaged. If the solar telescope is coaxially mounted with another telescope, ensure that it has its lens cap firmly attached.

Open the dome and point the telescope at the Sun. Many telescope control programs will not let you perform this action. Consequently, you may have to drive the telescope to position manually; please check with your observatory staff whether you need to do this, and if so, how it is done. Start your webcam, and using the shadow of the telescope, nudge the telescope until you see the disk of the Sun. At this point, you may need to focus the solar telescope and adjust the webcam's exposure time (or frame rate), brightness, and gain in order to get a clear image.

Once you are happy with the image, capture approximately 15 s of video. Lucky image this video file using Registax, for example, to produce a high-resolution image. Save this image for processing later and then shut down the observatory.

13.4.4 Analysis

Using the images obtained, measure the movement rates of sunspots over the face of the Sun over the course of 14 days.

Compare the images in hydrogen-alpha to those in white light and monitor how the sunspots and other transient phenomena, such as prominences, change over the course of the observations.

Most FITS manipulation software can draw a profile, for example, ATP. Use this feature to determine the brightness gradient of the Sun across its surface.

As with previous practicals, you should write up your report in standard laboratory style. You should discuss the different structures you see and compare how their appearance differs between hydrogen-alpha and white light. You should determine the rotation rate of the Sun and comment on how, if at all, it changes with latitude. The brightness gradient over the surface of the Sun is known as limb darkening. You should investigate what is causing this effect and whether it is time variable.

13.5   Eclipse Observing

It is a remarkable coincidence that the ratio between the diameters of the Moon and the Sun and the ratio of the distances from Earth to the Sun and the Moon are almost identical, meaning that both the Sun and the Moon appear to have an angular size of half a degree. The orbit of the Moon is inclined by 5.145 deg to the ecliptic, resulting in the apparent paths of the Moon and the Sun intersecting at two points, known as **nodes**. When the Sun and Moon are both positioned at opposite nodes, a lunar eclipse occurs, whereby the Earth's shadow falls onto the Moon, leaving it illuminated by only the light refracted through the Earth's atmosphere. This gives the Moon a red appearance, leading in recent years to the popular but misleading term "blood moon."

When the Sun and the Moon are at the same node, a solar eclipse occurs. Due to the smaller size of the Moon's shadow when compared to the Earth's, solar eclipses are less common at any one location, are briefer, and can be seen over a wider area of the Earth than solar eclipses.

Solar eclipses come in four broad types: A total eclipse occurs when the shadow of the Moon completely obscures the Sun, allowing the much fainter solar corona to be visible. This is visible to all observers in the **umbra** section of the Moon's shadow. Outside of the umbra, within the penumbra, observers will see a partial solar eclipse. This can usually be seen from a larger part of the Earth than the total eclipse. However, some eclipses can be observed only as a partial eclipse, because the umbra never intersects the Earth's surface. Partial eclipses, although more common at any one location, are considerably less impressive than total eclipses. As the Moon's orbit around the Earth is imperfectly circular, the apparent diameter of the moon can change by about 6%. When the Moon is at its apparent smallest diameter, it is not large enough to cover the Sun. This can result in an annular eclipse, which occurs when the Sun and Moon occupy the same node, but the apparent size of the Moon is smaller than that of the Sun. This results in the Sun appearing as a very bright ring, or annulus, surrounding the eclipsing disk of the Moon. An observer must be in the **antumbar** in order to see the annulus. Much rarer is the hybrid, or annular/total, eclipse caused by the eclipse transitioning from total to annular as the Moon's shadow tracks across the Earth's surface.

Solar eclipses are divided into four phases, the start of each being marked by a specific event. **First contact** occurs in a solar eclipse when the leading edge of the Moon makes contact with the Sun, which is when the observer enters the penumbra. **Second contact** occurs when the trailing edge of the Moon contacts the edge of the Sun. This occurs when the observer enters the umbra for total eclipses or the antumbar in the case of annular eclipses. Third contact occurs in a total or annular solar eclipse when the leading edge of the Moon touches the far edge of the Sun and totality ends. This is effectively the point when an observer reenters the umbra. The eclipse ends with **Fourth contact**, occurring when the trailing edge of the Moon is no longer in contact with the Sun and the eclipse ends. This is the point at which the observer leaves the penumbra.

Safety is paramount when you are undertaking observations of solar eclipses. There is **no** point during a partial or annular eclipse at which it is safe to observe without using either a solar filter or projection. The only period during which a total solar eclipse can be observed without a solar filter or projection is during the brief period of totality.

During an eclipse, if you are not using an instrument, you should always use specialist eclipse glasses from a reliable supplier. During the American 2012 solar eclipse, there were problems with online suppliers selling eclipse glasses that did not cut out the harmful UV from the Sun and therefore endangered the user's eyesight. Unfortunately, this put off many users from looking at the eclipse, even though they had purchased safe solar eclipse glasses. Before using eclipse glasses, just as with any solar filter, you should check for damage. You should do this by looking at the glasses for signs of damage and shining a bright light through them. The amount of light allowed to pass through them should be uniform, with no bright spots or scratches. If there is even a hint that the glasses are damaged, they should be destroyed to ensure that they are not used. Even a small hole could cause irreparable damage to the retina.

During the eclipse, you may wish to take images using a camera. The same rules apply to cameras as for the human eye. Always use a solar filter during the entire eclipse, except for the period of totality for a total solar eclipse. Failure to do so can cause catastrophic damage to the camera and if viewed through a DSLR, eye damage.

As a general rule of thumb, it is always best to set an alarm to go off with 30 s of warning of the end of totality so that everyone can get their eyes and equipment suitably protected before totality ends (Fig. 13.8).

There is very little "science" to be achieved during a partial eclipse, as most of the desired effects are not seen until at least 90% totality is achieved. However, in recent years there has been much interest in the temperature drop that has been reported during eclipses, and therefore you may wish to measure the local temperature during the eclipse.

During a total eclipse, there are a number of features that can be observed and recorded that are not available at other times, such as the **corona**, the outer atmosphere of the Sun.

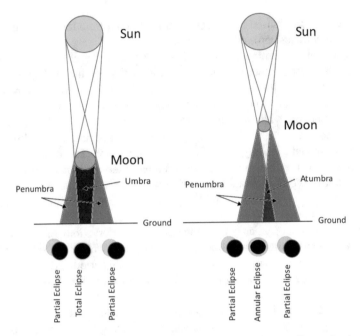

Fig. 13.8 Diagram of the position of the observer during a solar eclipse: reality to what is observed

The corona, as seen in Fig. 13.9, is, as the name suggests, a crown of plasma. It surrounds the Sun and is visible only during the totality of a solar eclipse or when you use a coronagraph. Spectroscopy measurements of the the corona show a temperature in excess of 1,000,000 K, which is much hotter than the surface of the Sun. The cause of this discrepancy was still unknown at the time of writing. The asymmetrical structure of the corona during eclipses is of interest, as it must in some way be reflecting the activity of the Sun and the activity of its magnetic field, and it appears to be linked to the Sunspot cycle. Solar cycle 24 began in 2010, peaking in 2014 with a lower level of activity compared to the previous cycles. The 2017 solar eclipse, shown in Figs. 13.9 and 13.10, occurred chronologically halfway between maximum and the likely end of cycle 24, appearing somewhat asymmetrical. Future observations will be of interest as we head into cycle 25.

From a scientific point of view, **Baily's beads** (Fig. 13.10) are of little interest, as they are fully understood and are caused by sunlight shining down lunar canyons and giving the effect of bright beads at the edge of the Sun. Given that the Moon has been mapped in detail by the Lunar Reconnaissance Orbiter (LRO), an observer should be able to identify the lunar geological features that caused a particular bead. Students who find themselves lucky enough to observe a total solar eclipse may find this an interesting project.

Perhaps one of the greatest mysteries still associated with solar eclipses is the existence of **shadow bands**. These appear as wavy lines of alternating light and

Fig. 13.9 Image of a total solar eclipse showing the corona. (Image courtesy of NASA/Gopalswamy)

dark that can be seen moving and undulating in parallel on plain-coloured surfaces, occurring before and after totality. There has been a number of explanations for the cause of this banding, but exact answers fail to materialise, perhaps hampered by the difficulty of imaging the effect. The most likely explanation is that it is a scintillation effect from the atmosphere, which becomes visible when the Sun is reduced to a point light source, which would occur just before and after totality, the same time at which the shadows are observed. It is analogous to stellar scintillation, but with the effect reversed in direction. The effect seems to have different levels of intensity between eclipses, with no shadow banding seen at some. Whatever is happening, using a pale surface to capture shadow banding during a total eclipse will contribute to our understanding of the effect and the atmosphere.

Given that there is a maximum of seven solar eclipses occurring in any one year and any one site seeing only one total solar eclipse in a typical human lifetime, planning is important. You will likely be travelling to see a total eclipse. Fortunately, solar eclipses can be predicted hundreds of years in advance, which will give you plenty of time to select a location on the path of totality and book flights and accommodation. I have seen multiple eclipses, including three total solar eclipses, and I generally plan two years in advance. Generally, it is best to stay away from urban areas and national parks. Not that they are bad observing locations, just that everyone else will think they are a good location, and you will also spend a considerable amount of time in traffic trying to get away from the site after the eclipse and will have to pay significantly more for accommodation.

Perhaps most of all, you should not observe a solar eclipse alone, especially if observing somewhere remote. Part of this is safety; it isn't going to go completely dark, but accidents do happen. Most of all, experiencing a total solar eclipse is something you should do with other people. Watch eclipses with friends!

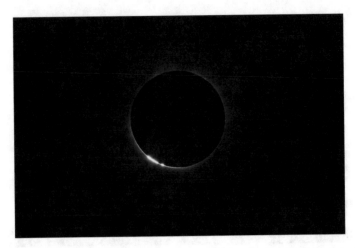

Fig. 13.10 Image of a total solar eclipse showing the Baily's beads effect. (Image courtesy of NASA/Aubrey Gemignaniy)

For choosing and planning for the eclipse, NASA has an extremely good eclipse site https://eclipse.gsfc.nasa.gov/solar.html, which shows eclipse type, date, and line of totality. It is also worth looking at the typical climate for the region of totality to reduce the chance of poor weather, although this does not prevent a single cloud from magically appearing right over the Sun at the moment of totality. Remember to adjust contact times for the local time zone and the offset due to the longitude and latitude of your website.

One last note: some observers become so focused on their equipment that they fail to take a few moments to look at the eclipse with their own eyes. Total solar eclipses are by far best observed with a mark one eyeball.

Chapter 14
Spectrography

Abstract In this chapter, we introduce the concept of spectrography. Building on the previous chapters on the nature of light and astronomical imaging, we discuss the spectrograph and show through a simple practical how to take, reduce, and interpret our first astronomical spectrum.

14.1 Introduction

Spectrography is one of the principal methods through which astronomers investigate the universe. The existence of dark absorption lines and bright emission lines set against the broad continuum emission and their unique association with energy changes, as discussed in Sect. 2.2, has enabled astronomers to to accomplish the following things:

- From the shift of a line relative to its normal position we can determine velocity and redshift and cosmological distance as well as infer the existence of exoplanets.
- From the width and strength of the lines we can determine relative abundances of elements and molecules as well as the temperature of the body.
- From the shape of the line, we can determine the pressures, density, and rotational characteristics of the body and infer the presence and strength of its magnetic field.
- The measurement of the continuum part of the spectrum, that part without lines, enables us to observe nonquantised emission processes and to determine the spectral energy distribution of the body.

Hence, spectrography provides us with a set of tools that allow us to determine a wide range of physical characteristics.

© Springer Nature Switzerland AG 2020
M. Gallaway, *An Introduction to Observational Astrophysics*,
Undergraduate Lecture Notes in Physics,
https://doi.org/10.1007/978-3-030-43551-6_14

14.2   The Spectrograph

In order to obtain spectra, we must disperse the light from the observed object such that photons of the same wavelength fall in the same location in our imaging device. There are two ways of doing this: by refraction using a prism and by diffraction using a diffraction grating. There is a third device that uses both methods, known as a grism. The grism is a prism with a grating etched on its surface.

It is also possible to undertake limited spectrography using narrowband filters. As these are tuned to a single spectral line, such as [OIII] or [SII], it is possible to measure the intensity of this line, although not its width.

The diffraction grating is a reflective or transmissive surface etched with regular lines, often over a thousand lines per millimetre. The lines in effect turn the source into multiple sources, one per line. These constructively and destructively interfere, creating multiple rainbow patterns on the detector.

Equation (14.1) shows the diffraction equation, where d is the line spacing in the grating, θ is the emergence angle, n is an integer known as the order, and λ is the wavelength:

$$d(\sin \theta) = n\lambda. \tag{14.1}$$

As we can see when $n = 0$, the dispersion angle θ becomes zero. The 0th-order image, therefore, is not a strictly a spectrum, but rather the image of the object. To either side of the object is a series of repeated spectra, becoming fainter as we move away from the 0th order. The first of these spectra, on either side of the 0th order, are the first-order spectra, which are typically the spectra that will be analysed. For reflection gratings, (14.1) becomes

$$d(\sin \theta_i + \sin \theta_e) = n\lambda, \tag{14.2}$$

where the term $\sin \theta$ is replaced by the sum of the sines of the incident angle θ_i and the angle of refraction, θ_e.

Figure 14.1 shows the layout of a traditional long-slit spectrometer as found at many small observatories. Light enters the spectrograph through a slit that is designed both to improve the spectral resolution and to isolate the source. Often, the size of the slit can be adjusted. Once inside the spectrograph, the light is collimated, so that the incident angle is consistent. The reflection grating disperses the light depending on wavelength. Typically, the grating can be rotated so that the area of the spectrum of interest will fall on the centre of the CCD. Likewise, some gratings are double backed with the option of a high- or low-resolution grating. The dispersed light passes through the camera's optical system and forms a spectrum on the CCD. Most spectrographs will have a slit illuminator to allow the alignment of the source and slit. They may also have a calibrated low-intensity light source such as a small argon or neon bulb. These will produce a strong reference line in the spectra for calibration purposes. Of course, if you are observing in a light-polluted area, the sodium doublet lines are ideal for calibration.

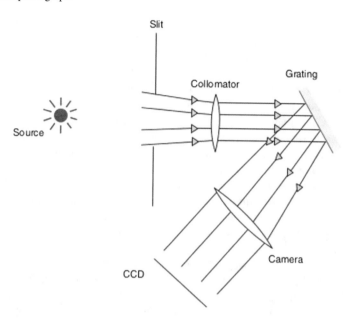

Fig. 14.1 Diagram of a long-slit spectrometer showing the key components of the slit, grating and detector. It is common for there to be a number of additional optical components inside of spectrometers that improve image quality

The spectral resolution of your spectrograph depends on the level of dispersion and the pixel size of the CCD in addition to the grating and the order being observed. Equation 14.3 determines the spectrograph's spectral resolution $\Delta\lambda$ from the grating width d, the angle of refraction θ_e, the order number n, the pixel scale x, and the focal length of the instrument F:

$$\Delta\lambda = \frac{d\cos\theta_e\Delta x}{nF}. \tag{14.3}$$

The ratio between the central wavelength λ and the wavelength covered by each pixel is the spectrograph's resolution, R. Hence, larger R indicates higher resolution. For example, the commercially available Skelyah Lhires III has an R value of 18,000, which equates to a resolution of 0.035 nm near the hydrogen-alpha line, which equates to a Doppler shift of $16\,\mathrm{kms^{-1}}$; good enough to observe the rotation of stars.

$$\frac{\lambda}{\Delta\lambda} = R \tag{14.4}$$

With the light dispersed over multiple orders and light lost in both reflection and refraction from optical surfaces, it should come as no surprise that a considerable portion of the light entering the spectrograph is not detected. A good surface will lose about 2% of the light reflected or passed, while the grating may lose 75%, and the CCD will be on the order of 80% efficient. Hence, in a typical eight-element configuration, total throughput will be on the order of $0.98^8 \times 0.75 \times 0.8 = 0.51$. So we lose about half our light within the optics. It is important, therefore, when undertaking photometry that the exposure time be long enough to get a good signal-to-noise ratio and that the slit be opened to the optimal aperture.

As with imaging, there is always the option of binning to improve the signal-to-noise ratio. As individual columns of pixels all represent the same wavelength, assuming that the image is not rotated, we can, and certainly should, vertically bin our spectra to a single row. If we have higher spectral resolution than we need, we also have the option to bin linearly after binning the columns. Binning is good practice if you are not particularly interested in the line width. It will also enhance the bright lines when the spectra are plotted and reduce confusion within the plot.

Just as it is possible to undertake field photometry, it is possible to undertake field spectrography, but not with a standard long-slit spectrograph as shown above. Multislit photometry is possible using a modified long-slit spectrograph. However, there are problems with this approach, and it can be inefficient.

Fibre-fed multi-object spectrographs (MOS) use a mounting plate with holes drilled where the image of the stars would fall. Integral field units (IFU) divide the field into a grid with a small lens covering each cell. The lenses focus light into fibre-optic cables glued to the back of the lens. The fibres feed into the spectrograph and form multiple spectra. IFUs are powerful tools for determining large-scale movement such as the rotation of galaxies.

It is possible to undertake slitless spectrography using an inline transmission grating such as a Paton Hawksley Star Analyser. In-line transmission gratings can be fitted to the nose of a camera or webcam or inserted into a filter wheel. They produce low-resolution spectra of all the objects in the field, which may result in confused and overlapping spectra. However, they are inexpensive, easy to use, and are becoming increasingly popular with amateur astronomers.

14.3   Taking and Reducing Spectra

To some degree, the taking and reducing of spectra is very similar to that of normal imaging. Although the software used for the reduction is different, in general, the camera control software is the same.

As with imaging, you will need to take dark, flat, and bias calibration frames and apply them either with the camera control software or a spectral reduction package such as RSpec, VSpec, IRIS, or IRAF. Remember that the spectrograph is a CCD, and it needs to be cooled and managed in the same way as a standard imaging CCD.

For each spectrum, both reference and calibration spectra are required; this is in addition to bias, dark, and flat frames.

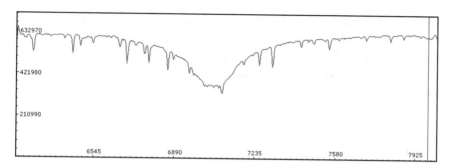

Fig. 14.2 Stellar line spectrum. Referenced but not calibrated for instrument spectral response. The x-axis is in angstroms, while the y-axis is in counts

The calibration will be of a bright source, with a known, standard spectrum and at the same air mass as the source; where possible, it should be of the same spectral type and the same exposure time. If the primary source being imaged has a known standard spectrum, there is no need for a calibration frame. The reference frame is a spectrum of a known source, perhaps an arc lamp or an LED, either within the spectrograph itself or external to it. The reference frame is used to convert pixel position to wavelength.

Once the spectra are taken, they should be turned into science frames by the subtraction of the bias and dark frames and the division by the flat frame. The spectra should be located within the frames and cropped out. A region of the sky should also be cropped out parallel to the spectra, with the same size and the same starting column. These sky regions contain unwanted lines from the atmosphere (as do the source spectra), and subtracting them from the source removes them.

The reference spectra (which have lines of known wavelength) are used to correlate pixel position to wavelength and the wavelength scale. When these are applied to the source and calibration spectra, they convert to line spectra plotting intensity against wavelength. This should appear somewhat like the stellar spectrum shown in Fig. 14.2.

Accounting for spectral response is more involved. The calibration spectra must be divided by the standard for that object. This process leaves a spectral response curve for your instrument setup. It will, however, be messy, due to the spectral lines. Something smoother is preferred, which can be achieved by fitting a polynomial, the instrument response curve. Inverting this curve and multiplying it by the source line spectra yields a calibrated spectra.

Normal practice is to normalise the spectrum by identifying the highest peak and dividing the entire spectrum by peak value.

Once a spectrum is reduced, spectral lines may be identified from their position along the curve.

14.4   Practical 10: Methane in the Atmosphere of Neptune

The solar system bodies Uranus and Neptune contain a significant amount of methane in their atmosphere. This methane is sufficient to be detected by a small instrument with a spectrograph.

14.4.1   Aims

You will be taking spectra of either Uranus or Neptune to identify key components in their atmospheres. You will learn how to operate a telescope fitted with a spectrograph and how to take, reduce, and interpret spectra.

14.4.2   Preparation

Before arriving at the site, you should have selected a suitable target and calibration star. You should have produced finder charts for the night and have sourced a standard spectrum. Uranus and Neptune are large enough to have a visible disk that should make identification straightforward.

14.4.3   Imaging

On arrival, open the dome and cool the camera(s). If needed, take flats or obtain suitable library flats.

Once the camera has been cooled to operating temperature, slew the telescope to a bright star and do a focusing check. If you have a flip mirror or off-axis instrument, use this to point the telescope and confirm that it is aligned and pointed at the correct piece of sky. When you focus the spectrograph, you may have to open the slit wide. Once you are aligned and focused, move the telescope to the source. Align the target in the slit, which should now be closed to a suitable width. Image the target, with binning if needed, making sure that every pixel stays within the linear range of the CCD.

Once you have suitable spectra, take dark and bias frames and then turn on the reference source and take another spectrum. Ensure that you can see the lamp lines in the image.

Once you have your reference, slew to your calibration source and align in the slit. Take the spectra, again while remaining in the linear range, and follow this with dark and bias frames. Then take spectra with the reference lamp on.

Save all your files to a transportable medium and check that they are in an acceptable condition, for example, that they have no plane trails, and once you are happy with your work, shut down the observatory.

14.4.4 Reduction

Once you have the spectra, you have to reduce them in order to get science line spectra:

- Using your favourite reduction package, remove the bias and dark frames from the spectra and divide by the flat.
- Rotate the frames so that the spectra are flat and not tilted. If the spectra are curved, remove this.
- Crop the spectrum and save to file. Perform the same action for the sky area, ensuring that the cropped size is the same as the spectra and that it starts at the same pixel column.
- Subtract the sky from the source to produce a sky removed spectrum.
- Load the reference spectra into a spectral reduction package such as IRIS, Vspec, or Rspec and bin them, resulting in a line plot of intensity versus pixel number.
- Locate the emission lines in your reference spectra, noting the pixel value for each line and its corresponding wavelength. Use this to find the linear dispersal of the spectrograph.
- Load the calibration spectrum and calibrate it using the pixel location of the wavelengths found in the reference spectra and the calculated linear dispersal. Applying this should create a line plot of intensity versus wavelength.
- Load the standard spectrum and bin it down to a line spectrum if it is not already, and divide the calibration spectra by the standard spectra.
- Fit a polynomial to the resulting spectra. It may be necessary to remove some spectral lines to get a smooth fit. Save this plot.
- Load the source spectrum and bin it. Calibrate this spectrum using the data from the reference spectra, as previously.
- Take the calibrated source spectrum and divide it by the polynomial you fitted to the calibration spectrum.
- Normalise your spectrum by finding the value of the highest peak and dividing through by that value.

14.4.5 Analysis

Using a spectral line catalogue, identify and label the spectral line within your spectrum. Pay particular attention to lines redder than 600 nm, as these may be molecular lines. Additionally, compare the solar spectrum to your spectrum.

Write up your observations and reduction in the standard format. You should discuss how the spectrum from your source compares to the solar spectrum and what this implies. If you saw molecular lines, you should discuss what this suggests about the atmospheric content of the planet's atmosphere and temperature.

Chapter 15
Radio Astronomy

Abstract Radio astronomy is becoming increasingly important in astrophysics, with increasingly complex, sensitive, and large instruments coming online. In the chapter, we introduce some of the concepts in radio telescope design. As teaching radio telescopes are still rare, we introduce radio astronomy with three practicals. Mapping the Milky Way, detecting meteors, and observing Jupiter, the last two of which can be undertaken with low-cost radio setups.

15.1 Introduction

In the early 1930s, Karl Jansky (after whom the jansky is named) was working on improving shortwave radio reception. Using a directional antenna, he noticed a signal that appeared and disappeared over a 24 h cycle. Using his equipment as a primitive radio telescope, he was able to determine that the source of the signal was not the Sun (although the Sun is a bright radio source), as he had previously suspected, but something in Sagittarius. We now know that this source is the supermassive black hole and radio source, located at the centre of the Milky Way, known as Sgr A*. So began the age of radio astronomy, and the number of known astronomical radio sources rapidly increased.

In the early 1960s, two physicists, Arno Penzias and Robert Wilson, were working on the Bell Lab's horn antenna in New Jersey (interestingly, Jansky was also working at Bell Labs when he discovered Sgr A). Penzias and Wilson were working on cryogenically cooled microwave receivers and were troubled by a continuous background signal being received. The signal was coming from all parts of the sky and was uniform in intensity. After a complete examination of the entire instrument, including removing a white dielectric deposit from within the horn (i.e., bird droppings), they concluded that the signal was not an artifact, but something real.

The physicists contacted astrophysicist Robert Dicke, thinking he might be able to help with the source of the mysterious signal. Dicke was, as it turns out, an inspired choice. He was working on the theory of the evolution of the universe, the Big Bang, and the prediction that there should be a remnant signal from the age of deionisation,

© Springer Nature Switzerland AG 2020

M. Gallaway, *An Introduction to Observational Astrophysics*,

Undergraduate Lecture Notes in Physics,

https://doi.org/10.1007/978-3-030-43551-6_15

seen as an isotropic microwave signal. Dicke even went as far as to suggest that the Bell Labs Horn antenna must be the perfect instrument to detect the signal. This was the very signal detected by Penzias and Wilson—the cosmic microwave background (CMB).

Penzias and Wilson were awarded the 1978 Nobel Prize in Physics for the discovery of the CMB. When you look at a detuned terrestrial television, about 1% of the signal you see is from the CMB. You are seeing the birth cry of the universe on your television.

A few years after Penzias and Wilson's discovery, a young graduate astronomer, Jocelyn Bell Burnell, was working on a dipole radio telescope in Cambridge, England, known as the hop field, due to its similarity to the supports used for growing hops.

Bell Burnell noticed a strange signal in her data. It was astronomical, strong and varying very rapidly and with a very consistent period. Initial fears that the signal was of alien origin were soon put aside when it was realised that Bell Burnell had detected the first pulsar, a dense rotating neutron star emitting two tight beams of radio waves that briefly become visible to the observer as the star rotates, very much in the manner of a lighthouse. Neutron stars are formed in explosions of supernovae, and this object was located in the heart of the Crab Nebula.

Bell Burnell's doctoral supervisors were awarded the Nobel Prize in Physics for Bell Burnell's discovery, though Bell Burnell was not, somewhat controversially. Bell Burnell is on record as stating that she believes that this was the right choice and that her supervisors were fundamentally responsible for the project. However, Bell Burnell has gone on to be a very successful and well-respected astrophysicist and is a strong advocate for the involvement of women in science.

Radio astronomy has become increasingly attractive to astrophysicists, as it can operate in daylight and most weathers (depending on frequency), is mostly unaffected by interstellar reddening, and can observe both high- and low-energy phenomena that are not accessible by conventional optical and infrared observations.

The principal limitations of radio astronomy are its general insensitivity and poor resolution. However, the facility with which multiple radio telescopes may be combined in an array, together synthesising a single large virtual telescope, of diameter equal to the greatest separation between component telescopes of the array (interferometry), have overcome these obstacles.

In the coming years, three very large interferometers will reach full operation. Both ALMA, a very large submillimetre interferometer, and LOFAR, a very large long-wave interferometer, are now operating. Work is about to start in South Africa and Australia on the Square Kilometre Array, and its completion will usher in a new era of radio astronomy.

15.2 Instrument Background

A radio telescope consists of two components: the antennas that collect the radio signal and the hardware that turns it into an oscillating electrical signal, with the frequency of oscillation being the same as that of the signal. This change occurs due to the electromagnetic characteristics of light, with the magnetic field inducing a current in the conducting antenna.

This current propagates down coaxial cabling, the transmission line, to the receiver, which measures the voltage being received, which is proportional to the signal strength and can therefore be taken as a measure of how loud the signal is. Within the transmission line, there will often be one or more low-noise amplifiers, designed to enhance the signal on its way to the receiver.

All antennas have five defining properties: impedance, directional characteristics, forward gain, polarisation, and beam pattern.

The impedance is the measure of the opposition that the antenna presents to the induced current. Impedance should be matched throughout the system, as mismatching can cause internal reflections and signal loss.

Some antennas, such as the one on your Wi-Fi router, are omnidirectional. Although some radio telescopes use omnidirectional antennas, for example LOFAR, it is generally not a desirable property of a radio telescope antenna. Hence, most radio telescope antennas are designed to receive, to the extent possible, radio waves from only one direction.

The forward gain indicates just how efficient the antenna is when pointed to a source as compared to a standard dipole antenna. It is measured in dB. A 3 dB forward gain means that your antenna is twice as sensitive as a standard antenna, while 6 dB gain would indicate an antenna that was four times as sensitive. Similarly, the backwards gain is the amount of signal that is lost as it leaves the antenna. Basically, with regard to forward gain, the bigger the better.

As we have previously discussed, electromagnetic waves consist of an electric wave at right angles to a magnetic wave. If the orientation of the electric field is always the same for all photons from a source, the source is said to be polarised. In effect, most astronomical sources have only limited levels of polarisation, although most antennas are polarised. Hence, half the photons are lost. This feature can be used to block unwanted terrestrial sources.

The beam pattern is the spread of radiation that would be caused if we used our antenna to transmit rather than receive. Do not worry if this seems an odd concept. Trust me—the mathematics works. We find that there is a main beam extending out from the antenna related to its field of view. However, there will be some other beams, known as side lobes, extending as pairs on either side of the main beam. You should be aware that you can detect a source in a side lobe, but the side lobe will have lower sensitivity.

The most common antenna is the dipole, which consists of two lengths of wire, tube, or rods that are linked to the radio receiver via the transmission line. The most common dipole is the half wave, which, as its name suggests, has antennas half the length of the wavelength we wish to receive. The dipole antenna has a doughnut-shaped beam pattern.

The Yagi-Uda antenna, now more commonly known as just a Yagi, is a multi-dipole antenna, again normally half wavelength. Yagi antennas can be optimised to balance the competing forces of gain versus bandwidth. Hence in terrestrial applications, such as receiving television, which requires wide bandwidth, Yagi antennas tend to have reflectors to improve gain. As anybody who has tried to mount a new television antenna will know, a Yagi tends to have a small beam pattern, which is good for astronomical applications. Hence, the Yagi antenna is often used in low-cost systems.

At the more professional end of radio astronomy, we come to the parabolic reflector, called simply a "dish." These are like satellite receiving dishes only scaled up. In fact, domestic Sky Television dishes have become popular amateur radio astronomy antennas. Traditional large-dish radio telescopes such as the 4.5 m R.W. Forrest teaching telescope at the Bayfordbury Observatory (Fig. 15.1), consist of a large dish designed to focus radio waves to a point. At this point is located a simple, quarter-wavelength waveguide (just a conductive tube, in effect) designed to direct the signal to the receiver. The waveguide is surrounded by the feed horn (or can), which is a metal can open at one end that guides the incoming radiation onto the waveguide and protects it from stray radiation.

The receiver attached to your radio telescope will have an operating frequency range and a centre frequency. Your receiver, if designed for radio astronomy, should be able to perform two functions. The first is a continuum plot. In this mode, it will just display the variation in source intensity over time. It should also be capable of spectrography. In this mode, it moves the frequency that it is listening to in small steps, plotting the intensity of the signal against wavelength or frequency. In many cases, it is possible to adjust the integration time, i.e., how long is spent at each wavelength, and the step size, the range over which it will be listening to each point along with the scan. In effect, increasing the step size is like binning with a CCD. Your receiver will most likely report the signal strength in volts. With some difficulty, this can be turned into instrument flux, and then standard flux. However, just as with counts with CCDs, this conversion is not always required.

Radio sources can be divided into two basic classes: low and high energy. Lower-energy sources are quantised sources with low energy levels. Typical of this is the 21 cm line. This is a strong emission line for monatomic hydrogen caused by the change in the quantum spin state of the proton and electron. Monatomic hydrogen (HI) is the most common form of hydrogen and hence is a vital tool for the identification and mapping of galactic structure. Other examples of radio spectra lines are ammonia, water, methanol, and carbon monoxide. All these lines occur at low temperatures and low energies. High-energy sources include the Sun, Jupiter, black holes, active galaxies, pulsars, and supernova remnants. In these cases, strong magnetic fields accelerate charged particles in a spiral, causing the emission of braking radiation with a characteristic spectrum related to the Larmor radius. Such sources are often hot and are also X-ray emitters, whence the odd effect that X-ray astronomers also tend to be radio astronomers.

Fig. 15.1 The 4.5 m R.W. Forrest teaching radio telescope located at the Bayfordbury Observatory UK. (Image courtesy of the University of Hertfordshire)

The bandwidth limitations of your receiver and antenna setup will determine what you will be able to observe with the instrument. The 21 cm line is one of several regions of the radio spectrum that are protected. This is a strong emission line for monatomic hydrogen caused by the change in the quantum spin state of the proton and electron. Monatomic hydrogen (HI) is the most common form of hydrogen and hence is a vital tool in the identification and mapping of galactic structure. However, you will not be able to observe pulsars using a telescope.

15.3 Radio Practicals

The following radio practicals are dependent on the correct equipment. The first of these, Mapping the Milky Way, requires a telescope capable of receiving 21 cm emission. The second, Counting Meteors, requires the simple setup of a yagi antenna and an FM receiver. A copy of Spectrum Lab would also be very useful.

The last practical requires a RadioJove setup, or something similar that can pick up emissions at 8–40 MHz from a pointlike source. Constructing such an instrument is good practice in itself.

15.3.1   Mapping the Milky Way at 21 cm

The presence of dust and gas in the plane of the galaxy makes it impossible to map our galaxy globally by optical astronomy. The 21 cm line emission from monatomic hydrogen can, however, be used. From the Doppler shift and intensity of the line, we can determine the kinematics and size of the hydrogen clouds.

Figure 15.2 shows our galaxy from above the disc, with the spiral arms marked. As we observe through the galactic plane, any line of sight may intersect hydrogen clouds at different distances from the galactic centre. We determine these distances

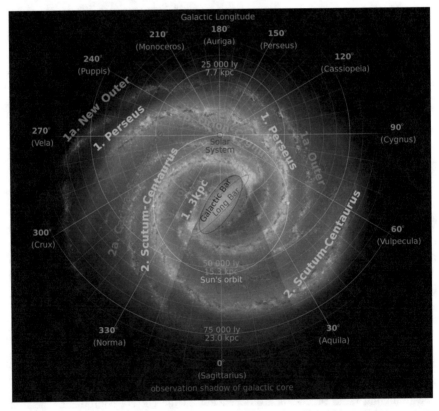

Fig. 15.2   Image of the Milky Way as derived from observations. (Courtesy NASA and R. Hurt)

using the cloud velocities formula as derived from the shift in their 21 cm line and using a model of the rotation of the clouds around the galactic centre.

You should look at your planetarium package with the Milky Way turned on, to determine what parts of the galaxy you will be able to see and where in the sky it is in galactic coordinates. As the telescope effectively has only one pixel, you will be scanning the Milky Way by taking a spectrum and then moving the telescope half a beam width before taking the next spectrum. This will be needed for the whole visible region, so it is best to plan how to do this in advance.

Aims

The principal aim of the practical is to give the user experience in using a radio telescope and taking data. However, the observer will also be determining the rotation curve of the Milky Way and mapping the HI distribution.

Observation

Once at the telescope, start it up and slew to a bright radio source, for example Cas A or the Sun. Once there, take spectra with the radio telescope and confirm that you are seeing a signal. You might have to adjust the gain or integration time to see detail in the spectra.

Once you are happy with the operation of the telescope, slew to the first region of the Milky Way. Once on target, take the spectra. Note the velocity position of the peak in the curve and its intensity as well as its position in the sky in galactic coordinates. If there is more than one peak, you might be seeing two or more clouds, at differing distances, in which case note them all. Save the plot to a suitable medium.

Slew the telescope half a beam width to the next region of the Milky Way that you are going to examine and take other spectra. Do this until you have covered at least a 20° section of the galaxy.

You might wish to take multiple spectra of each section if you have time, as this will help identify your uncertainties.

At the end of the observation run do not forget to park and shut down the telescope.

Analysis and Write-Up

We now wish to map the HI in the galaxy and find the galaxy's HI rotation curve. The local standard of rest is the origin of our reference system, i.e., the current position of the Earth compared to the galactic centre at any one time. However, the Earth is moving in relation to the local standard of rest as it orbits the galactic centre. Hence, any Doppler shifted signal we receive from an HI cloud contains this movement as well as that of the cloud around the galactic centre. Figure 15.3 shows the relationship between the galactic centre (GC), the radial velocity of the Sun relative to the local

Fig. 15.3 Illustration of the Sun's movement around the galactic centre, the position of an HI cloud and local standard of rest

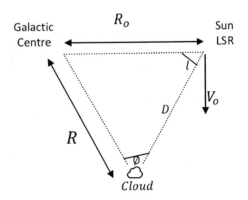

standard of rest V_o, the distance from the Sun to the galactic centre R_o, the distance of the cloud from the galactic centre r_l, and the cloud's radial velocity relative to the galactic centre.

The value of R_o is known from observation to be 8.5 kpc, and V_o to be 232 km^{-1}. The line of sight velocity of the cloud, V, as measured by your radio telescope, is given by

$$V = \left(V_R \frac{R_o}{R} - V_o \right) \sin(l), \tag{15.1}$$

where l is the galactic longitude. We assume that latitude has little or no influence.

We can see from (15.1) that for any given line of sight, where l is fixed, V depends only on $(V_R \frac{R_o}{R})$. If the rotation curve is flat or decreases with the radius, the maximum value for V should be achieved when R is at its smallest value, and this occurs at the point where the angle between R and the line of sight is 90°. Hence, the maximum line of sight velocity $V_m(l)$ is given by

$$V_M(l) = V_c(R_L) - V_o \sin(l), \tag{15.2}$$

although this applies only to $-90° < l < +90°$.

We now have only to find R, which can be done using

$$R = \frac{V_R R_o}{\frac{V}{\sin(l)} + V_o}. \tag{15.3}$$

For each spectrum, using the largest line of sight velocity, calculate R_L/R_o and $V_c(R_L)$; then make a plot of $V_c(R_L)$ against R_L/R_o, remembering to use error bars.

Using this rotation curve, you can now find the distance to all the clouds you observed by solving the quadratic equation $R^2 = D^2 + R_o^2 + 2DR_o \cos(l)$.

Using the distance to each cloud and its galactic longitude, you should make a map of the Milky Way.

Write up your report in the standard style, including as usual all your data and calculations (and errors) as well as the all-important rotation curve and map.

If the orbit of the clouds around the galactic centre is Keplerian, as we would expect, V_c will scale by the square root of R. Is this what you see in your findings, and if it isn't, what do you think the implications of this are? You should discuss this in your write-up.

Your map should show the HI structure of the Milky Way. You should discuss in your write-up how your map compares to Fig. 15.2 and what you think are the implications of any difference.

15.3.2 Radio Meteor Counting

There is a vast number of meteoroids in the solar system. Most are the size of a grain of salt, some the size of bricks, and a very few are as big as a car or larger.

When they enter the Earth's atmosphere, as about million kilograms of meteors do a year, they ionize the atmosphere as they decelerate from about $20\,\mathrm{kms^{-1}}$. This creates a bright streak of light in the sky at an altitude of about 80 km—a meteor. A few of these objects survive the fiery passage through the atmosphere and are found as meteorites.

There is a seasonality to the appearance of meteors throughout the year. Periods may have few meteors or many, sometimes hundreds an hour, an event known as a meteor shower. Showers are associated with comets due to the trails of dust they produce as they orbit the Sun, with the showers occurring where the Earth's orbit intersects the orbit of the comet. A meteor unassociated with a shower is known as a sporadic.

Of course, meteors arrive day and night and in all weathers, so most go unnoticed by everyday observers. However, there is a relatively inexpensive and easy way to detect them at all times and in all weathers.

The key to this is the ionization of the atmosphere by the incoming meteoroid. Ionized gas is reflective to short-wavelength radio waves. So by placing such a radio source over the horizon and pointing an antenna and suitable receiver at the source, we find that we can detect the ionization trail. Normally, we receive no signal from the source, since it is over the horizon. However, when a meteor event occurs between the source and the observer, the ionized air acts as a mirror that allows us, very briefly, to see the source. Hence, we hear a meteor in the receiver as a short blip, normally Doppler as the meteor moves in the radial direction. This blip often sounds like a woohoo from a fairground ride ghost train.

Fortunately, we don't have to create our own sources. VHF television masts and military space surveillance radars are very suitable sources, although, of course, this was not the designers' intention (Fig. 15.4).

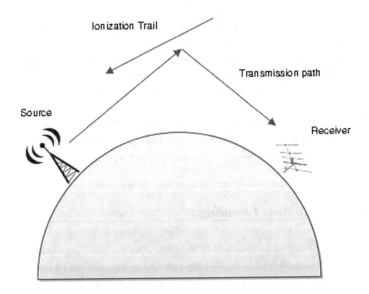

Fig. 15.4 Illustration of the detection of meteors using radio reflection

Preparation

You will need to observe over a period of 24 h. You can either set up the system to record the process or take shifts. You will need to identify the transmitter you are using, its frequency, and location relative to the observatory. You will also have to confirm that there is a meteor shower at the time you are observing and where the radiant is. If you are using a program like Spectrum Lab to record the data, you should ensure that you are familiar with its operation.

Observations

On arriving, you will set up for observations if that has not been done already. This means pointing the antenna in the correct direction and tuning the receiver to the required frequency. If you are using Spectrum Lab, you will need to ensure that the sound output of the receiver is plugged into the line-in port sound card of the computer. Check to make sure you are picking up meteor reflection. If you are not using Spectrum Lab or something similar, I suggest that the sound be recorded for backup.

Over a period of 24 h, note down every meteor trace and the time it occurred. Long notes or streaks turn out to be satellite traces (or a bad alternator in the security detail's car, as happened once to the author) and can be ignored. You should note every 15 min the number of detections in that 15 min window.

Analysis and Write-Up

Plot the 15 min count against time. The curve should not be flat. Discuss what you think this means and the physical process that could cause this. What do you think the implication of the radiant on the completeness of your detections is and how might it be improved? What are your sources of error and how significant do you think these are?

Write up your report in a standard format including your plot and data.

15.3.3 Radio Emission from Jupiter

In 1955, a burst of radio emission was detected from Jupiter at 22 MHz, the first planet to be detected in the radio spectrum. Part of this emission is synchrotron radiation from particles trapped in Jupiter's rather powerful magnetic field. This component is at 178–5,000 MHz.

A second radio component at 3.5–39.5 MHz was seen that appears to be strongly related to the orbit of Jupiter's inner moon, Io. It is thought that Io and Jupiter are linked by a flux tube, a series of magnetic lines. Electrons spiral down the tube, emitting radio waves in the process. The spot at which the electrons descend into Jupiter's atmosphere has even been detected from Earth. Figure 15.5 illustrates this interaction.

NASA has developed RadioJove, a school project designed to introduce the public to radio astronomy.[1] The radioJove project includes a receiver kit, software, and a simple antenna. This should be suitable for detecting radio emission from Jupiter. However, in order to get a higher signal-to-noise ratio, it might be worth investing in a suitable Yagi, a tracking mount, and an off-the-shelf receiver.

However, independent of the kit you have access to, as long as you have a setup able to receive decimetre emission from Jupiter, you should be able to complete this practical.

Aims

Using a simple radio telescope and receiver, you will observing the radio emission from Jupiter and interpret the variation in signal strength.

Observations

You need to observe Jupiter for a number of days (or nights) in order to get a complete understanding. If you have only a loop antenna, then observing during daylight may

---

[1] The RadioJove project can be found at http://radiojove.gsfc.nasa.gov/.

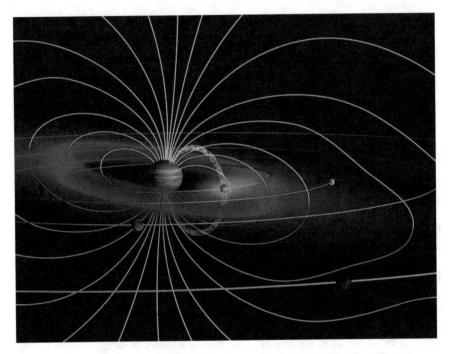

Fig. 15.5 Illustration of the magnetic interaction between Jupiter and Io. The flux tube can be seen in *green*. (Image courtesy of John Spencer. Copyright John Spencer CC BY-SA 3.0)

be difficult if the Sun is very active. If you have a directional antenna that can track, then daylight observations are possible, but better results can be obtained after dark.

Turn on the receiver, setting it to the correct frequency. If necessary, point the antenna at Jupiter and start tracking. There are short-term and transitory variations in the radio emission from Jupiter. You are looking for more long-term variation. To find this, record the signal strength every 5 min, and then uses the average of 12 signal strengths to find the mean hourly signal. Use this also to calculate your uncertainties.

Write-Up and Analysis

Once you have all your data, plot time against intensity, remembering to use error bars. Using your favourite planetarium program, plot the latitude of Io, Europa, Callisto, and Ganymede against time. Compare these three plots with your radio data.

Write up your results in the standard format including your data, plots, uncertainties, and any calculations. Discuss the nature of the variation in the radio signal and its relationship to the main Jovian moons.

Chapter 16
Astrometry

Abstract Astrometry deals with the position and movement of astronomical objects in the sky. It is one of the most ancient applications of astronomy, and in its modern form, it is used to determine to the position of stars within the galaxy and the orbits of comets and asteroids, and to detect the existence of exoplanets. We show how to perform astrometry using SAO DS9 to determine the proper motion of an asteroid, and from that its distance from the Sun.

16.1 Introduction

Astrometry deals with the position and the movement of astrophysical bodies like the planets, asteroids, comets, and stars. The production of an astronomical catalogue, with the positions of those objects, requires astrometric observations and calculations.

Astrometry is the oldest scientific application of astronomy, as it was the astronomical measurements of Tycho Brahe that allowed Kepler to develop his three laws of planetary of motion and Newton to discover his law of universal gravitation. It also gave the first hint that Newton was incorrect, leading to Einstein developing general and special relativity, which we first confirmed through observation of stars during an eclipse.

In modern astronomy, astrometry helped in the discovery of the supermassive black hole in the galactic centre as well as, indirectly, in the detection of exoplanets.

Astrometry is one of the techniques we can use to determine the distances to the stars. The method of stellar parallax requires very high precision astrometry to measure the angular displacement of a nearby star when referenced to more distant stars, caused by the rotation of the Earth around the Sun. A 1 arcsec displacement is equal to a distance of 1 pc. However, parallax-related changes for a source beyond 100 pc are difficult to detect. Although there has been some progress with very long baseline radio interferometry that has positional resolutions of sub-milliarcseconds, this pushes direct distance measurements out to kiloparsecs for radio sources, some of which, for example masers, are associated with other features such as star formation.

© Springer Nature Switzerland AG 2020
M. Gallaway, *An Introduction to Observational Astrophysics*,
Undergraduate Lecture Notes in Physics,
https://doi.org/10.1007/978-3-030-43551-6_16

The European Space Agency's (ESA), Gaia space mission is attempting to measure the position of a billion stars with a positional accuracy of 24 milliarcseconds. If successful, this will be a major step forward in our goal of mapping the Milky Way. Astrometry is also used to determine **proper motion**, the movement of a star across the sphere of the sky—as opposed to radial motion, that is, toward or away from the Earth. For example, suppose a star's position is measured twice 10 years apart. The first position is (α_1, δ_1), and a decade later, it is at (α_2, δ_2). The changes in the angle are $\mu_\alpha = \alpha_2 - \alpha_1$ and $\mu_\gamma = \gamma_2 - \gamma_1$. To find the proper motion, we apply $\mu^2 = \mu_\gamma^2 + \mu_\alpha^2 \cdot \cos \gamma_1$.

Measuring changes in position is challenging over short time scales (i.e., years). The change in position is small, and the resolution of a ground-based telescope is limited by atmospheric seeing (which is why Gaia is space-borne). Finding the centre of a star can be accomplished by finding the centroid of the Gaussian distribution of the star, and repeating the process increases the reliability of the measurement. Likewise, using many stars of known position to locate the position of the unknown source also reduces uncertainties. Once the direction, position, and hence proper motion vector of the star is known, we may need to account for the proper motion of the Sun and Earth depending on the application.

We have already briefly mentioned plate solving in this book. Plate solving is a form of astrometry that relates pixel location to sky location. Plate solving programs such as Pinpoint determine the plate scale of your image from the pixel size and focal length of the image. It then identifies the stars within the field, at which point it uses one of several algorithms to match star patterns in its library to those in the image. Once it has a match, it identifies the coordinate-to-pixel transform and writes it into the header in the form of the world coordinate system WCS, which might include scaling, rotation, and field curvature corrections.

16.2   The Proper Motion of Asteroids

The identification of solar system objects, for example comets and asteroids, is undertaken by their observed orbital properties, which are determined from repeated observations over consecutive nights, if possible. Three nights is the minimum the Minor Planet Centre suggests.

In this practical, you will observe an asteroid at opposition over successive nights to determine its proper motion and approximate distance, assuming a circular Keplerian orbit.

16.2.1   Aims

You will be taking images of a bright asteroid at opposition over several evenings to determine its proper motion and its solar distance. You will also learn how to use the DS9 vector and blink function.

16.2.2 Planning

Using the Minor Planet Centre (MPC) website http://www.minorplanetcenter.net/iau/MPEph/MPEph.html, you should identify an asteroid at or near opposition that is bright enough and in the right position to be observed. You should download the ephemeris for this target to be sure that you have the position for the days and time you will be observing. Create finder charts for each night as usual.

16.2.3 Observing

- On each night of observing, cool the camera as usual and start up the telescope. As we are undertaking astrometry, we don't need dark or bias frames.
- Move the telescope to a bright star, focus up, and ensure that the pointing is good. Make any adjustments as needed.
- Move to the target (make sure you have the right RA and Dec for that night) and take a short image, using your finder chart to ensure that you are in the right location.
- Take several longer-exposure images to ensure that your stars in the image have Gaussian PSF.
- Plate solve the images, using either a local program such as Pinpoint or an online program such as astrometry.net.
- Save your images to a portable medium and shut down the observatory.

16.2.4 Reduction and Analysis

You will need to open your images in pairs is SAO DS9 with tiled frames and then blink the frame so you can identify the object.

- Open DS9 and from the frame menu select Tile Frame and then New Frame.
- Click on the left-hand frame, and from the file menu select Open and open your first image. Select the right-hand tile and load your second image into it.
- You will now have two FITS images displayed. On each tile, in turn, select Zoom Fit to Window and then scale 99.5%.
- From the frame menu select frame match WCS. This will orient the two frames so that they match. You should have some like those displayed in Fig. 16.1.
- Select Frame Blink, and the two frames will be displayed alternately every half second. Your object should be seen moving against the background stars.
- Make a note of its location in each frame. Repeat the procedure for all your frames until you have located the asteroid in all your images.

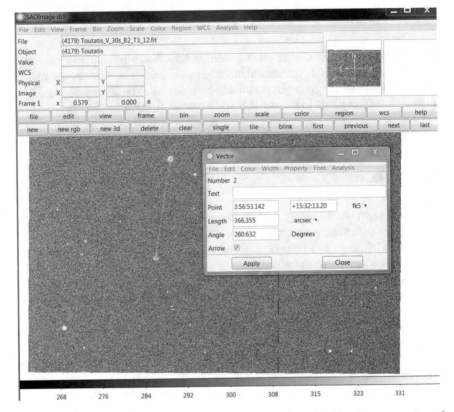

Fig. 16.1 Screen shot of two frames loaded into SAO DS9 prior to blinking. (Images courtesy of University of Hertfordshire and the Smithsonian Astrophysical Observatory)

You now need to determine the position of the asteroid in each image. You could do this by moving the pointer over the asteroid and reading off the WCS coordinates. However, that is not very accurate.

- There should be some bright stars in the field whose positions you know. It is best to have this in decimal degrees, as working in sexagesimal makes the math challenging.
- From the Region menu choose Shape and then Vector. Then draw a vector from the reference star to the asteroid.
- Click on the vector to activate it and select Region Centroid. This step is very important, as it determines the centre of the PSF for both the reference star and the asteroid.
- Double click on the vector to bring up its information window. You should note the vector's length in arc seconds and its angle.
- Add vectors from at least three reference stars and use the vectors to determine the position of the asteroid. Use the mean position, and use the range as your error.

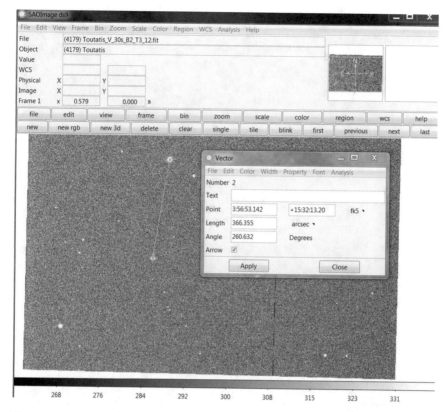

Fig. 16.2 Screen shot of a vector between an asteroid and a reference star and the vector information window. (Images courtesy of University of Hertfordshire and the Smithsonian Astrophysical Observatory)

- Now you know the initial and final positions of your asteroid, you use $\mu^2 = \mu_\gamma^2 + \mu_\alpha^2 \cdot \cos \gamma_1$ to determine the proper motion.

You should now be able to work out the distance of the asteroid from the Sun. Remember that the asteroid is at opposition, and so the velocity vectors for the Earth and the asteroid are parallel. Assume that both orbits are circular, Sun centred, and Keplerian (Fig. 16.2).

16.2.5 Write-Up

Write up your report in the standard style, including all your calculations and images.

Discuss how you think you could improve your uncertainties and reflect on how you would deal with an asteroid that was not at opposition.

Glossary

AB System   A photometric standard based on instrumentation.

Absolute Magnitude   The magnitude of a body at a distance of 10 pc for a non-solar-system object and a distance of 1 AU for an object inside the solar system.

Absorption Lines   Dark lines in a spectrum caused by the absorption of light by atoms and molecules between the emitting source and the observer.

Achromatic   A two-component lens designed to reduce chromatic aberration.

ADU Bit Number   The number of bits that the processor of an ADU uses. This limits the maximum number it can work with.

Airy Disc   The pattern produced by the diffraction of light within the optics of a telescope or other such device.

All-Sky Photometry   A photometric technique that does not use reference stars.

Angular Resolution   The telescope's ability to separate two objects, given by the average wavelength of the light being observed divided by the aperture of the telescope.

Aperture   Either the diameter of an optical system such as a telescope, or the size of the region being measured in photometry.

Aperture Photometer Tool   A freeware multiplatform application designed to perform aperture photometry.

Aperture Photometry   Photometry using an aperture around an image of a star with the count within the aperture being integrated.

Apochromatic   A three-component lens designed to reduce chromatic aberration.

Apogee   For an object in an orbit, this is the point at which it is farthest from the object it is orbiting.

Apparent Brightness   The brightness of a source as viewed from Earth.

Apparent Magnitude   The magnitude of a source as viewed from Earth.

Apparent Solar Time   The local solar time based the observation of the Sun.

Arc Minute   A measure of angle. One 60th of a degree.

Arc Second   A measure of angle. One 60th of an arc minute.

© Springer Nature Switzerland AG 2020
M. Gallaway, *An Introduction to Observational Astrophysics*,
Undergraduate Lecture Notes in Physics,
https://doi.org/10.1007/978-3-030-43551-6

Array Size   The size of a CCD or CMOS in pixels.

Astigmatism   A problem caused by the incorrect alignment of optical components.

AU   The mean distance between the Sun and Earth, 149,597,870,700 m. Used mainly for working with objects within the solar system.

Background   In photometry, the unwanted signal arising not from the source, but from the sky, internal sources of noise, etc.

Bahtinov Mask   A mask that causes a distinctive pattern when the telescope is in focus.

Balmer Series   A series of hydrogen lines caused by the transition of an electron down to the $n = 2$ energy level.

Band   A range of wavelengths. For example, the R-band is the wavelength over which a Johnson R-band filter is transparent.

Barlow   A device to increase the effective focal length of a telescope.

Bias   The signal within a CCD frame that comes from the read process and other such sources.

Bias Frame   A zero-second exposure image used to remove the bias.

Binning   The joining of pixels to form one large pixel.

Black Body   A perfect emitter and absorber of radiation.

Blooming   The spike effect caused by reading a saturated pixel.

Bolometric   Overall wavelengths. For example, the bolometric magnitude is the magnitude integrated over the entire spectrum.

Broadband Filter   A filter that covers multiple spectral lines.

Cassegrain   A telescope using a hyperbolic secondary mirror placed parallel to the parabolic primary, reflecting the light down the tube and through a hole in the centre of the primary with the focal plane lying just outside the tube at its rear.

Catadioptric   A telescope design that uses a correcting lens (known as the correcting plate) in front of the primary mirror.

CCD   Charge-coupled device. The light-sensitive part of a CCD camera.

Celestial Equator   The Earth's equator projected onto the sky.

Charge Diffusion   The loss of electrons during the CCD transfer process.

Chromatic Aberration   The appearance of coloured rings in refractors, due to different wavelengths being brought to different focal points.

Circularly Polarised   Polarised to the rotational orientation of the photon's electric field rather than its linear orientation.

Circumpolar   An object that is either visible to an observer all year round or never visible.

Cold Pixel   A pixel reporting low at all times.

Collimation   The alignment between the optical components in a reflecting telescope. A telescope out of collimation has elongated point sources.

Colour Index   The ratio of two fluxes in different bands. Colour is a distance-independent value.

Colour–Magnitude Diagram (CMD)   A plot of magnitude versus colour (for example, B versus B-R).

Coma A problem with a reflector whereby light incident to the edge of the mirror is brought to a slightly different focus from light at the centre, resulting in the image becoming smeared out.

Coordinated Universal Time A standard time based on GMT. The different between GMT and UT is currently minor.

Cosmic Ray A high-energy charged particle. The source of cosmic rays is not yet fully understood, although high-energy events such as supernovae are thought to be likely candidates.

Count A dimensionless instrument-specific unit, indirectly related to flux.

Dark Adaptation A combination of physiological responses that increase the ability to see in low light levels. Dark adaptation takes about 20 min to reach its optimal level. It is lost almost instantly when the eye is exposed to bright light.

Dark Current The noise caused by thermal electrons.

Dark Frame An image of the same exposure time as the light frame but with the shutter closed. Used to remove the dark current.

Diagonal An optical device that changes the pathway of light as it leaves the telescope.

Diffraction Halo A halo that appears around bright objects due to diffraction.

Diffraction Spike A crosslike spike appears around bright stars in images. Often added by amateurs in Photoshop.

Digitisation noise Noise caused by an ADU that is a digital device.

Dithering Moving the frame a few pixels between exposures so that some sources are not exposed to the same pixels.

Dome Flats Flat frames that use the inside of the dome as the illumination source.

Ecliptic The path along which the Sun appears to travel.

Elliptic Coordinates A coordinate system using the ecliptic as the fundamental plane.

Emission Lines Bright lines in a spectrum caused by the presence of atoms and molecules in the emitting source.

Emission Nebula Any nebula that emits light, normally due to embedded stars.

Equation of Time The translation between mean solar time and apparent solar time.

Equator The great circle on a sphere perpendicular to the meridian and equidistant from the poles.

Exoplanets Planets orbiting stars other than the Sun.

Extended Objects A resolved object, for example a planet or a galaxy.

Extinction Coefficient The level of extinction.

Eye Relief A measure of the comfort of an eyepiece.

Field Curvature A problem associated with the projection of a curved surface (the sky) onto a flat one (the CCD).

Field of View The amount of sky visible through the instrument, measured in degrees or arc minutes.

Field Stop The aperture of an eyepiece.

Filters A device designed to allow only a range of specific wavelengths to pass through to the detector.

FITS   Flexible image transport system. The standard format for astronomical images.

FITS Header   The part of a FITS image containing image data but not imaging data.

Flat Fielding   A process to deal with nonuniformity of a CCD.

Flat Frame   An image of the twilight sky or the inside of the dome. Used for removing the flat field noise.

Flip Mirror   A flat mirror that can be moved from a position in which the mirror is parallel to the telescope to one in which it is inclined by 45° to the telescope.

Flux   The amount of energy transferred through an area in a set time.

Focal Length   The distance between the first optical component and the first point of focus. A main characteristic of a telescope.

Focal Reducer   A device to decrease the effective focal length of a telescope.

Forbidden Line   A spectral line that cannot be produced on Earth due to the low gas densities needed.

Free-Bound Emission   Emission caused by the capture of a free electron. This contributes to continuum emission.

Full Well Capacity   The number of electrons a pixel can hold.

FWHM   The full width half maximum, a measure of the spread of a Gaussian profile such as seen with a star.

Gain   The ratio of photons to count in a pixel in a CCD.

Gravitational Reddening   The general relativity effect that causes the light from high-gravity environments to have an increased wavelength.

Great Circle   A circle on a sphere whose diameter intersects the centre of the sphere, so that the circle has the same circumference as the sphere.

Greenwich Mean Time   The mean solar time at the Royal Observatory in Greenwich.

Hertzsprung–Russell (HR) diagram   A plot of absolute magnitude against colour. HR diagrams describe stellar evolution and are one of the foundations of modern astrophysics.

Home   The zeroing position of a telescope.

Hot Pixel   A pixel reporting high at all times.

Hour Angle   Local sidereal time minus the RA of the object.

Hour Circle   Any great circle celestial sphere that passes through the celestial poles perpendicular to the equator.

Hubble Palette   The guidelines for how the narrow-band Hubble filter set is displayed in colour images.

Hydrogen-Alpha   A Balmer series emission line at 656.28 nm, caused by an $n = 3$ to $n = 2$ hydrogen transition.

Image Processing   The process of producing a high-quality image by manipulation of the data.

Image Reduction   The process of turning a light frame into a science frame.

Integrated Magnitude   The total amount of radiation in a set band coming from an object, expressed as a magnitude. Most often used for extended sources.

International Astronomical Union (IAU) The governing body for astronomy made up of professional astronomers from around the world.

Interstellar Reddening The reddening of light from an object, caused by the preferential scattering of short wavelengths over long ones.

IRAF A professional Linux data reduction package for astronomy.

Isotropically In every direction. Most stars radiate isotropically.

Jansky A unit of flux used in radio astronomy.

Julian Date A standard dating system using the day as the base unit with the origin at noon on 1 January 4713 BCE.

Lab Book A journal containing a complete description of all observations and experiments including calculations and observations. A good lab book should enable other investigators to reproduce the lab book's author's experiments.

Light Frame The unprocessed digital image of an astronomical object.

Light Pollution Any source of manmade stray light.

LIDAR An optical version of radar used in atmospheric physics and the detection of small objects and changes in distance.

Limb Darkening The effect whereby the edges of gaseous bodies such as stars appear to darken at the edge.

Linearly Polarised Polarised to the linear orientation of the photon's electric field rather than its rotational orientation.

Local Sidereal Time The sidereal time local to an observer.

Lucky Imaging A imaging technique using a high-speed video camera in order to produce high-resolution images.

Luminance Frame An image normal in clear, used to enhance detail in a colour image.

Luminosity The total amount of energy radiated by a body over a specific bandwidth. If calculated over the entire spectrum, it is known as the bolometric luminosity.

Lyman Series A series of hydrogen lines caused by the transition of an electron to a lower energy level.

Magnification The ratio of the angular size of an object seen through a telescope and seen with the naked eye.

Main Sequence The area of the HR diagram that represents stars that are core hydrogen burning.

Maksutov–Cassegrain A catadioptric that uses the correcting plates as the reverse of the secondary.

Master Frames A calibration frame made by combining multiple frames.

Mean Solar Time The mean time as measured by the Sun, whose principal unit is the day.

Meridian The great circle on a sphere perpendicular to the planes of the celestial equator and horizon.

Meridian Flip A problem with equatorially mounted telescopes that requires them to "flip" over when a source crosses the meridian.

Metallicity The ratio of hydrogen to another heavy element, normally iron, in a star.

Narrow-Band Filter   A filter that covers one spectral line.

Neutral Density Filter   A very broad filter that removes light equally over a very large range of wavelengths. These are normally used to cut down the light from a bright source such as the Sun or Moon.

Neutrino   A subatomic particle with zero charge and an extremely low mass. The neutrino interacts weakly with matter and is produced in vast numbers during nuclear synthesis.

Newtonian   A reflecting telescope in which a flat mirror is set at 45° to the primary mirror and just inside the focal point so that the point of focus is outside the tube, often close to the top.

Nodding   Going off target; for example, when you cannot get a representative background in the current image.

Parsec   The distance at which an object would show a parallax displacement of one second of arc. It equates to $3.08567758 \times 10^{16}$ m.

Pedestal   The charge added to a pixel to prevent wraparound during mathematical processes.

Perigee   For an object in an orbit, this is the point at which it is closest to the object it is orbiting.

Photoelectric Effect   An effect by which certain materials produce a current only when exposed to light of a certain wavelength.

Photometry   The science of measuring the brightness of an object.

Pipeline   A series of programmes used to process an image automatically.

Pixel Size   The size of an individual pixel.

Planck's Law   A law of physics that describes the relationship between emission and the temperature of a black body.

Planetary Nebula   A nebula formed by the ejection of the outer layers of a red giant.

Plate Solving   A process by which the WCS transform for an image is found.

Point Spread Function   In astronomy, how the light from a source spreads out over the CCD. For a star, this should be a Gaussian distribution.

Polar Aligned   In terms of telescope mounts, a mount one axis of which is aligned with the celestial poles.

Polarisation Filter   A filter that only lets pass photons whose electric field is orientated in a specific direction.

Population I   Young stars like the Sun.

Population II   Older stars such as found in globular clusters.

Population III   The very first stars. Yet to be detected.

Propagation of Uncertainties   The methods by which mathematical operations affect errors.

Proper Motion   The motion of an object against the background stars.

Proton–Proton Chain   The nuclear process chain that converts hydrogen to helium, resulting in the production of energy. There are four such chains, with a varying degree of importance within a star, dependent on its mass.

Quantum Efficiency   In relation to CCDs, the efficiency of the pixels at converting photons to electrons.

Quantum Mechanics The fundamental branch of physics that describes the universe at the atomic and subatomic length scales.

Rayleigh Criterion The non-Airy resolution of a telescope.

Read Noise Noise created by the process of reading the CCD.

Readout Noise Noise associated with the ADU.

Residual Bulk Image Ghost images caused by saturation in a previous image

SALSA J A freeware multiplatform application designed to perform time series photometry.

SAO DS9 A FITS manipulation package built by the Smithsonian Astronomical Observatory.

Saturation The point at which a pixel is full and can detect no further photons during an exposure.

Schmidt–Cassegrain A class of reflecting telescope based on the Ritchey and Chrétien design but with a correcting plate.

Science Frame A light frame that has had dark, bias, and flat frames applied.

Seeing The angular dispersion of stars caused by atmospheric turbulence.

Sexagesimal Notation The description of angles as hours, minutes, and seconds (or degrees, minutes, and seconds) as opposed to decimal degrees.

Shot Noise Noise that originates from the discrete nature of the source, for example from a photon or electron.

Sidereal Time Time as based on the Earth's rotation measured relative to the fixed stars rather than the Sun.

Signal-to-Noise Ratio The radio between the source signal and the noise. High signal-to-noise data has less uncertainty.

Sky Brightness The intensity of the light being emitted and reflection of the sky. The sky brightness is the main limiting factor in deep-sky observations.

Sky Flats Flat frames that use the twilight sky as the illumination source.

Small Circle A circle on the surface of a sphere whose diameter does not intersect the centre of the sphere, that is, a circle that is not a great circle.

Spectrum The rainbow-like distribution of light made by a spectrograph.

Spectrograph A device that splits light regions of similar wavelengths.

Spherical Aberration An optical problem associated with spherical mirrors. Light rays incident at different points on the mirror come to focus at different points. The Hubble Space Telescope famously suffered from spherical aberration.

Stopping Down Reducing the aperture of a telescope.

Summing Co-adding images.

Surface Brightness The magnitude per unit area.

Transit Telescope A telescope designed just to observe objects as they cross the meridian.

Transmission Window A region in the electromagnetic spectrum to which a body is transparent. This term typically refers to the Earth's atmosphere.

Vega System   A photometric standard based on the apparent magnitude of Vega in that band.

World Coordinate System   A system to convert pixel position to sky position.

Zero Point   A calibration number to convert an instrument magnitude of flux to a standard one. It normally changes from night to night.

Index

© Springer Nature Switzerland AG 2020
M. Gallaway, *An Introduction to Observational Astrophysics*,
Undergraduate Lecture Notes in Physics,
https://doi.org/10.1007/978-3-030-43551-6

Printed in the United States
By Bookmasters